IMAGES
of America

TOWNSEND

I0806216

On the Cover: William "Claude" Derris came into the Townsend community in the late 1930s or early 1940s. He purchased the old button-factory building from the Bashor family and turned it into a motel. Derris was very interested in photography and took many pictures of the Townsend area as well as the many visitors to his motel. He would take his visitors on tours of the vicinity and to the Great Smoky Mountains National Park. This picture is of Ernest "Frosty" and Mabel Hall Myers, who lived next door to the Derris Motel and were good friends with Derris. (Courtesy of the William Derris Slide Collection, MS. 2123. University of Tennessee Libraries, Knoxville, Special Collections.)

IMAGES
of America

TOWNSEND

Missy Tipton Green and Paulette Ledbetter

ARCADIA
PUBLISHING

ISBN 978-1-4671-1211-6

Published by Arcadia Publishing
Charleston, South Carolina

Printed in the United States of America

Library of Congress Control Number: 2013955975

For all general information, please contact Arcadia Publishing:
Telephone 843-853-2070
Fax 843-853-0044
E-mail sales@arcadiapublishing.com
For customer service and orders:
Toll-Free 1-888-313-2665

Visit us on the Internet at www.arcadiapublishing.com

It was very hard to think of to whom we could dedicate this book; there are so many people who helped tremendously with this process and so many who want to preserve the history of Tuckaleechee Cove and Townsend. Because of their continuous support for this project, we dedicate this book to the Myers family of Tuckaleechee Cove. This includes many people who helped research and provided photographs and wonderful stories. As we researched the Tuckaleechee Cove area, many of the families could be traced back to the Myers family. John and Jane Dunn Myers came into Tuckaleechee Cove from Greene County early on, and this was the origin of all the Myerses from Tuckaleechee Cove and Cades Cove.

Contents

ACKNOWLEDGMENTS

Presenting the story of Townsend has been a learning process for us. We did not realize there was so much history in the Tuckaleechee Cove area. We knew some of the history but have done extensive research and many personal interviews with different individuals who have a great knowledge about the area. There is no way we could have completed this project without the continuous help of Marilyn Myers Byrd. No matter what we asked her for, she was always willing to help out. We are truly honored to have her working with us and have enjoyed her wonderful stories. Shirley McNiell Heatherly and Barbara "Bobbie" McNiell Handley have also helped tremendously. Their wonderful knowledge of the Townsend area and Little River Lumber Company and Little River Railroad Company is outstanding. Their father, Stuart McNiell Sr., was one of the officials of the Little River Railroad & Lumber Company. They have many treasured photographs of the company, Townsend schools, and family members. Shirley, Bobbie, and Marilyn have become some of our dearest friends, and we are greatly indebted to them for all of their help in creating this book. Many other people have contributed pictures and information to us, including Bernard Myers, Eddie McClanahan, Vance Shuler (Joann Effler Shuler photographs), Louise Tipton Loan, Fred Lawson, Stephen Webber (John W. Oliver photographs), the Townsend Alumni Association, Dana Davis Lund, Charlie Myers, Johnnie Bryant Sparks, and Mary Thompson Headrick. Without these wonderful contributions of great photographs, there is no way a book like this could have been completed. We also obtained permission from the University of Tennessee Library, Special Collections, to use its collection of photographs by William "Claude" Derris, owner of the Derris Motel in Townsend. Other resources used were Great Smoky Mountains National Park photographs, the collection of William Cox Cochran's photographs at the University of Tennessee Library, one of the Albert G. "Dutch" Roth photographs, as well as our own collections. We are not perfect in our publication; if we have information that is not correct, we apologize. Many thanks to everyone who helped us in this endeavor. We hope we can help preserve the history of Tuckaleechee Cove and Townsend area.

INTRODUCTION

Situated in Tuckaleechee Cove, one of the several "limestone windows" on the northern base of the Smoky Mountains, is Townsend, Tennessee, also known as the "Peaceful Side of the Smokies." Little River, which starts high in the Smokies, meanders east to west through Tuckaleechee and drains most of the cove. Native Americans were the first inhabitants in the area, and archeological digs have been done in Tuckaleechee Cove with finds dating back to 2000 BC. It has been established that the Native Americans had a village near the northern entrance by 1200 AD. In about 1600, the Cherokee began arriving and building a series of villages along Little River. The name *Tuckaleechee* derives from the Cherokee name *Tikwalitsi*, though the original meaning is unknown. By the time of arrival of the first European Americans in the late 18th century, the Cherokee villages had been abandoned. Indian War Path was an alternate path of the Cherokee that followed the waters of Little River to the Tuckaleechee villages, where it led eventually through Indian Gap and on to the lower Cherokee settlements in South Carolina. A short route from the valley towns to the over-hill towns led through Ekanetelee Gap, and this alternate path was used in the settlement of the coves. Many North Carolina settlers came through Ekanetelee Gap into Cades Cove and Tuckaleechee Cove, especially from the Pennsylvania German settlements in Rowan County, North Carolina. The Virginia settlers probably came down the valley and worked back up the streams into the coves.

When the legislature of the Southwest Territory created Blount County from a portion of Knox County on July 11, 1795, the site of Tuckaleechee Cove was definitely outside the limits of legitimate settlement. When the first Blount County court met at Abraham Wear's house, John Walker was granted leave to build a public mill on his own land in Tuckaleechee beside a branch of Little River. The court also ordered that a public road be planned and laid out the nearest and best way from the mouth of Dry Fork (now the east boundary of Townsend) in Tuckaleechee to John Craig's mill on Pistol Creek. It seems that although Tuckaleechee Cove was out of bounds for legal settlement, landowners there would be considered part of Blount County.

This quote was from a geologist in 1869: "Tuckaleechee Cove lies just within Blount County and is separated from Wears Cove by a narrow neck, or ridge, the two being about a mile apart. In leaving the latter, we pass through a low gap and then descend perhaps 300 feet into Tuckaleechee. Little River flows through it, and in leaving it, cuts out a narrow pass through the mountain walls upon the west, thus forming the gateway of the imprisoned basin. Tuckaleechee has been settled for about as long as neighboring Wear's Cove and now contains nearly 100 families."

For many years before Blount County was officially established in 1795, Tuckaleechee Cove and the flatlands of the Little River valley were lush, green hunting regions for the Cherokees of the Great Smokies. Sometime during this era, pioneer Peter Snider came up from old Fort Gamble looking for a settlement site. He liked the cove lands and river valley, but the Indians at that time were not entirely peaceful, and Snider retreated, determined to come back another day. He did come back, and he built his cabin and made his settlement near the old junction to Dry Valley.

The first reference to a land grant in Tuckaleechee Cove is a North Carolina grant issued in 1791 for 1,000 acres to Charles McClung and James Wood Lackey. It was located in Greene County on Little River and adjoined a 2,000-acre survey that included the town of Tuckaleechee. There is no way of knowing who may have first settled on these lands, but family tradition suggests that Peter Snider was the first settler and that he was once driven out by the Indians. There is no definite evidence to support this, however. The first registered transfer of lands was by John Walker Sr. of Knox County, Territory South of the Ohio River to John Walker Jr., dated April 22, 1793. In 1797, a check of the Hawkins Line reported two intruders in Tuckaleechee Cove, which could have been John Walker and Peter Snider. The Hawkins Line, also known as the Meigs Line, is a survey line that resolved prior survey controversies between the Cherokee Nation and US territory in 1802. By the First Treaty of Tellico, on October 2, 1798, Tuckaleechee Cove became a legitimate territory for settlement. By the 1880s, the lumber industry was in full swing, probably because of two key innovations: the band saw and the logging railroad. In 1900, Col. W.B. Townsend of Pennsylvania purchased about 86,000 acres of virgin forests stretching from Tuckaleechee Cove to Clingmans Dome. He received a charter for his new company about one year later and named it Little River Lumber Company. Little River Lumber Company began to prosper and was among the largest commercial logging operations in southern Appalachia. Colonel Townsend saw a need to transport his lumber, so plans for a railroad were started. Construction of a railroad in nearby Walland, Tennessee (the home of the Schlosser Leather Company tannery), began around 1902. The railroad interchanged with the Knoxville and Augusta, predecessor of the Knoxville & Charleston Railroad. The eight miles from Walland to Townsend opened for operation in 1903, as did the three miles between Townsend and the forks of Little River. The coming of industrialization transformed the isolated farming community of Tuckaleechee Cove into a bustling mill town. The name was later changed to Townsend, after Col. W.B. Townsend, who brought in many industrial jobs to the people living there. E.J. Kinzel started a mountain retreat in 1894 in Tuckaleechee Cove, which in later years turned into a mountain hotel with two healing mineral springs. Kinzel Springs became a booming retreat where many families traveled to spend an entire summer. It offered a dance pavilion, tennis courts, croquet, swimming in nearby Little River, and visits to the healing spring waters.

One

The People of Tuckaleechee Cove and Townsend

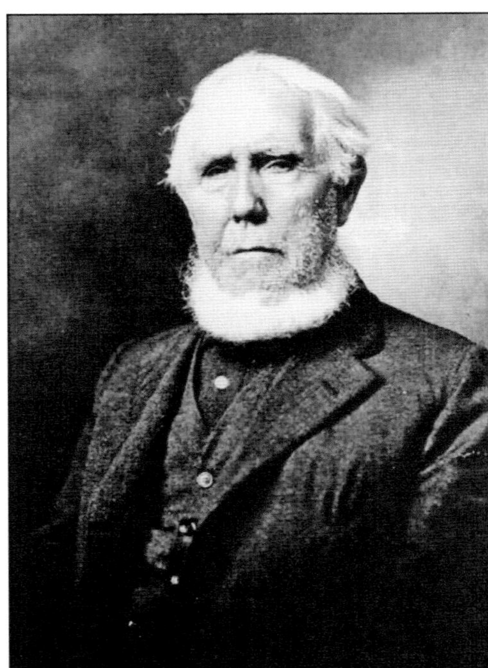

John "Sleepy" Myers was born on May 25, 1842, and died on January 26, 1923. On October 30, 1851, he married Margaret Peg Bird. Myers was the first postmaster in Tang, which is now called Townsend, and operated the post office out of his home from 1886 to 1906. The Myers family lived on a farm in the Tuckaleechee Cove area. He was engaged in many community and business interests. He was very well known and respected and was a great supporter of the church. In the early days, it was necessary for each family to be as nearly self-sufficient as possible. Myers, as well as every man, to a certain extent, was his own smith, cooper, cobbler, carpenter, and much more. (Image courtesy of *The Lure of the Great Smokies* by Robert Lindsay Mason.)

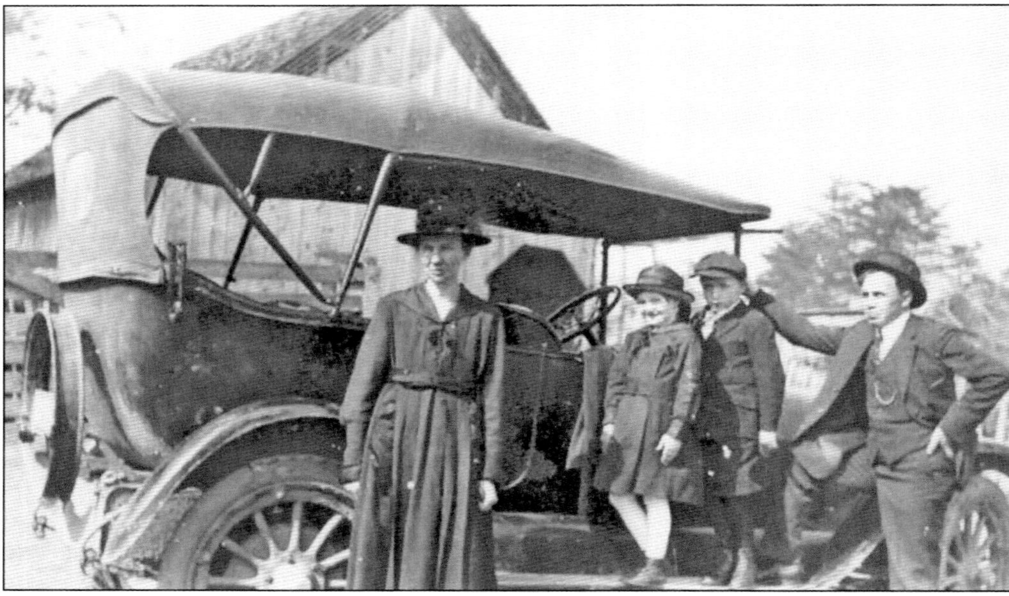

Pictured from left to right are Emma Scott Myers, Ruth Myers Pickle, Harold Joe "Beef" Myers, and Joe Myers. Joe (1883–1945) was the son of John "Sleepy" Myers and Margaret "Peg" Bird Myers. He was a well-known amateur photographer and salesman in East Tennessee, and he took many photographs of Townsend and Walland during the 1920s and 1930s. (Courtesy of Ralph Walker Jr.)

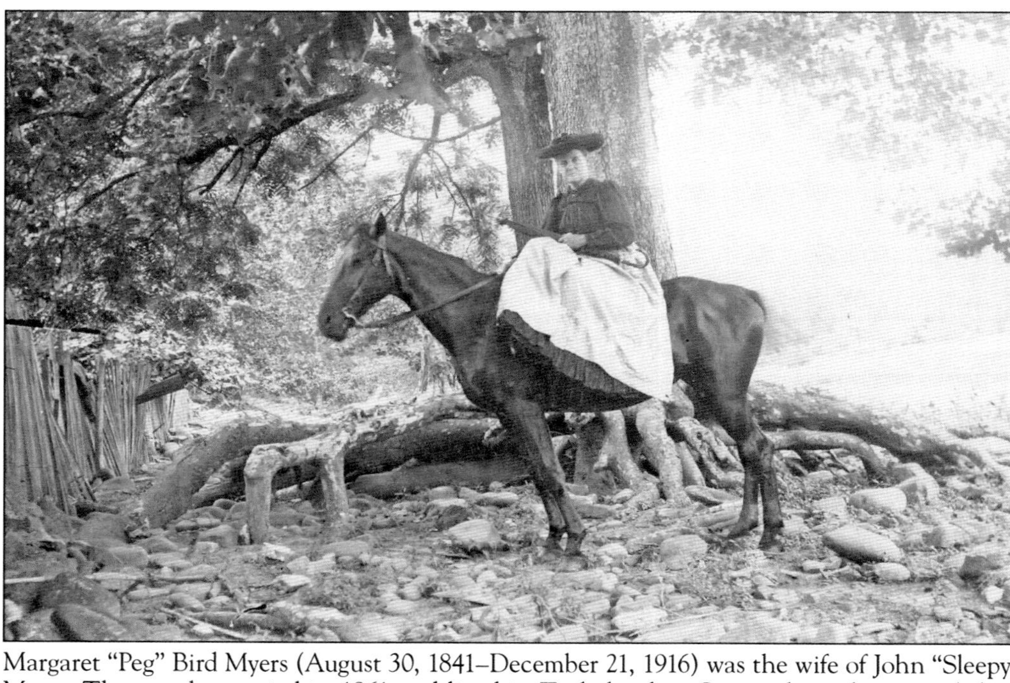

Margaret "Peg" Bird Myers (August 30, 1841–December 21, 1916) was the wife of John "Sleepy" Myers. The couple married in 1861 and lived in Tuckaleechee Cove, where they raised their family. (Courtesy of Ralph Walker Jr.)

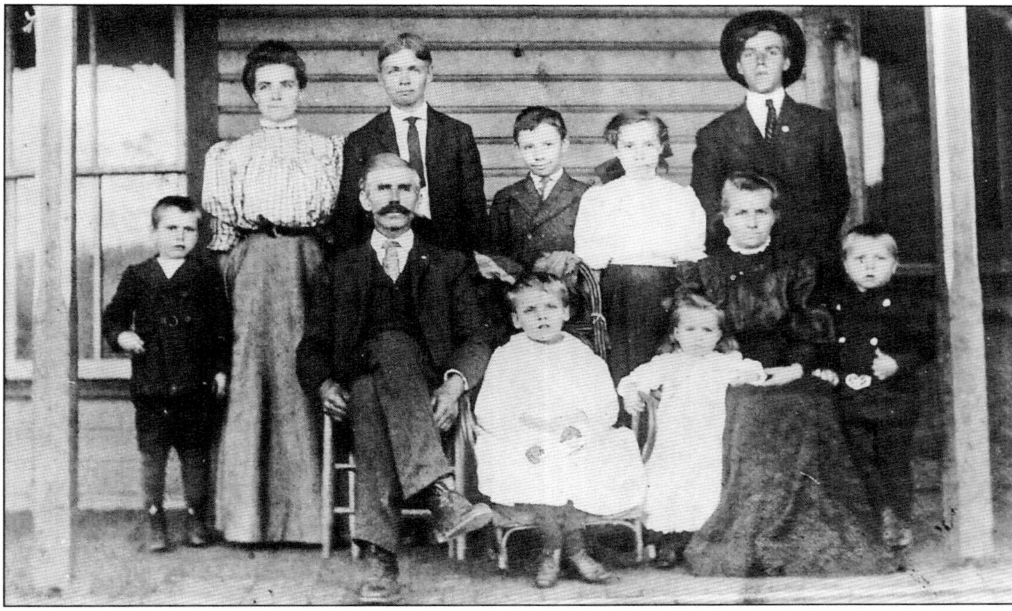

Pictured is the family of Joseph and Laura Feezell Myers. They are, from left to right, (first row) Joseph, Glenn, Stella, Laura, and Leonard; (second row) Lee, Mary, Bruce, Earl, Blanche and Pruitt. Joseph and Laura inherited a portion of the William Myers farm. Laura raised three rows of cotton for stuffing quilts and making clothing for the children. Using a carder, she cleaned and removed the burrs from raw cotton. (Courtesy of Marilyn Myers Byrd Photograph Collection.)

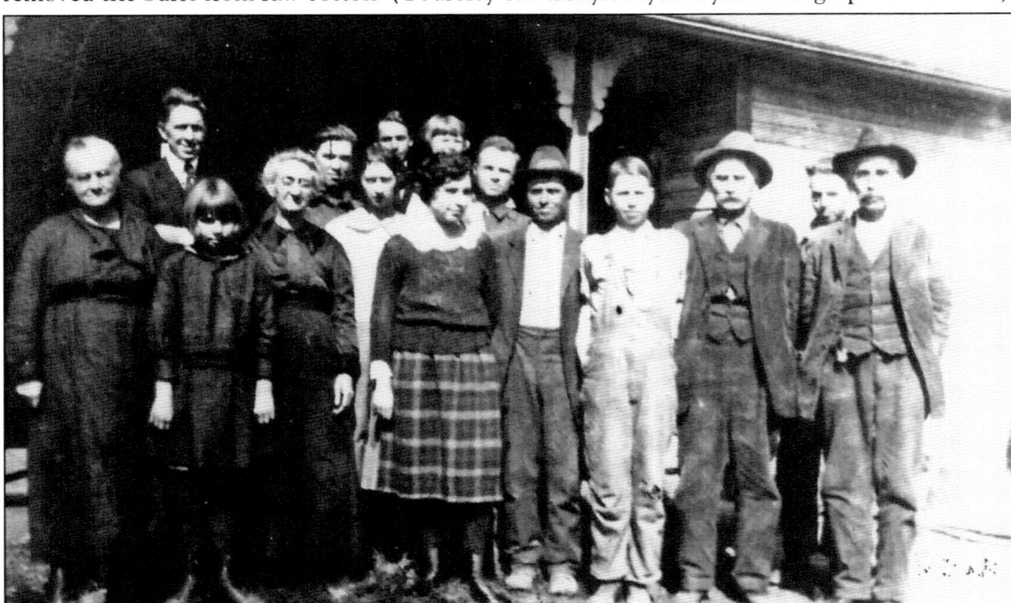

This is the family of Joseph R. Myers c. 1922. They are, from left to right, (first row) Laura Myers, Mildred Myers, Mary Myers (Sherman's wife), Louise Myers (Bruce), Stella Myers, Will "Pot Leg" Myers, Ross Myers, Sherman Myers (Mary's husband), and Joseph R. Myers; (second row) Dan Elmer, Otis Tipton, Lee Myers, Mildred Frances Myers on father Bruce Myers's back, and Leonard Myers. (Courtesy of Marilyn Myers Byrd Photograph Collection.)

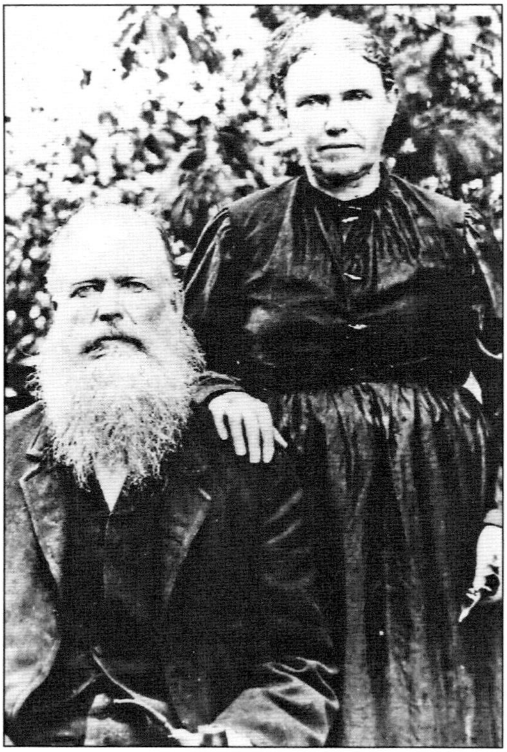

Rufus and Rhoda Emert Law are pictured here. Rufus was the son of Abraham Law of the White Oak Sinks area near Cades Cove. The Laws were devout Methodists. (Courtesy of Marilyn Myers Byrd Photograph Collection.)

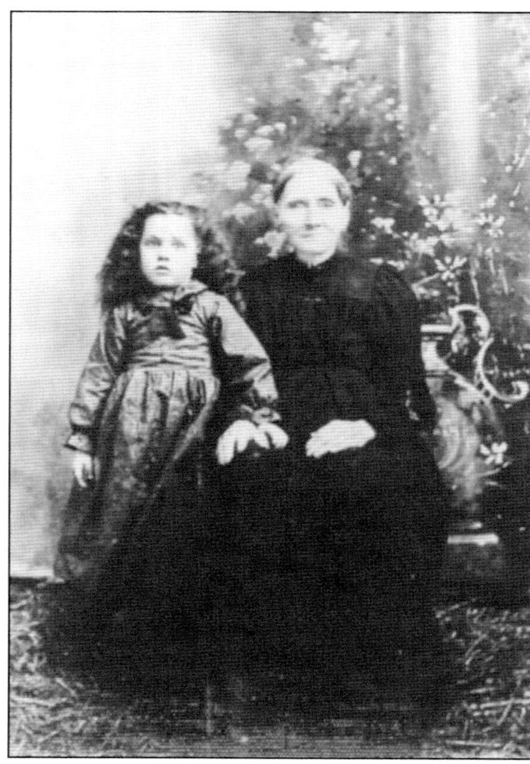

Lizzie Walker poses with her great-great-grandmother Rhoda Emert Law. After Lizzie's parents died, she was raised by the Headrick family. (Courtesy of Marilyn Myers Byrd Photograph Collection.)

William Hurst Dunn and Dorothy C. Snider Dunn are seen here. Dorothy was the granddaughter of Peter Snider and Mary Montgomery Snider. Peter is thought to have been one of the earliest settlers in Tuckaleechee Cove, and according to tradition, his family was living at Fort Henry, and Mary Montgomery's, at Fort Gamble, when the pair married. (Courtesy of Missy Tipton Green.)

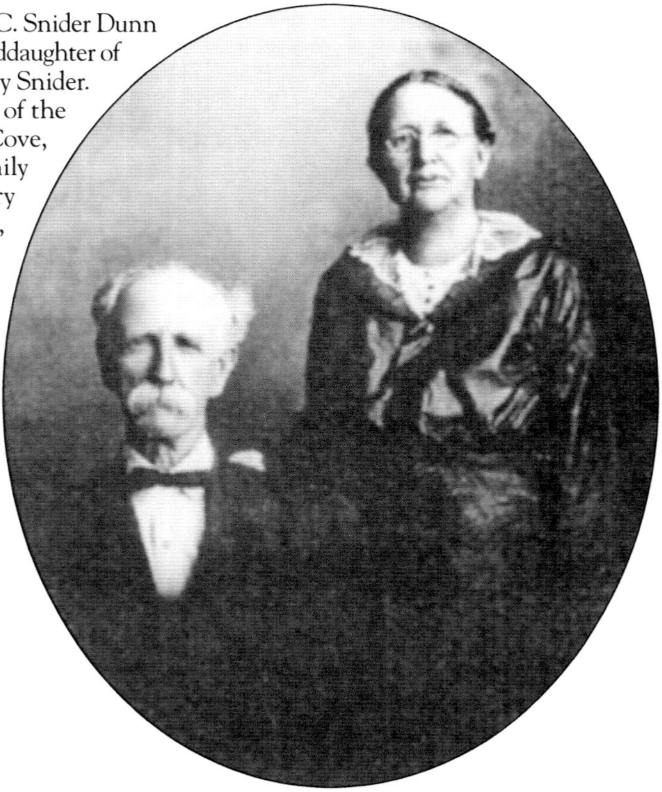

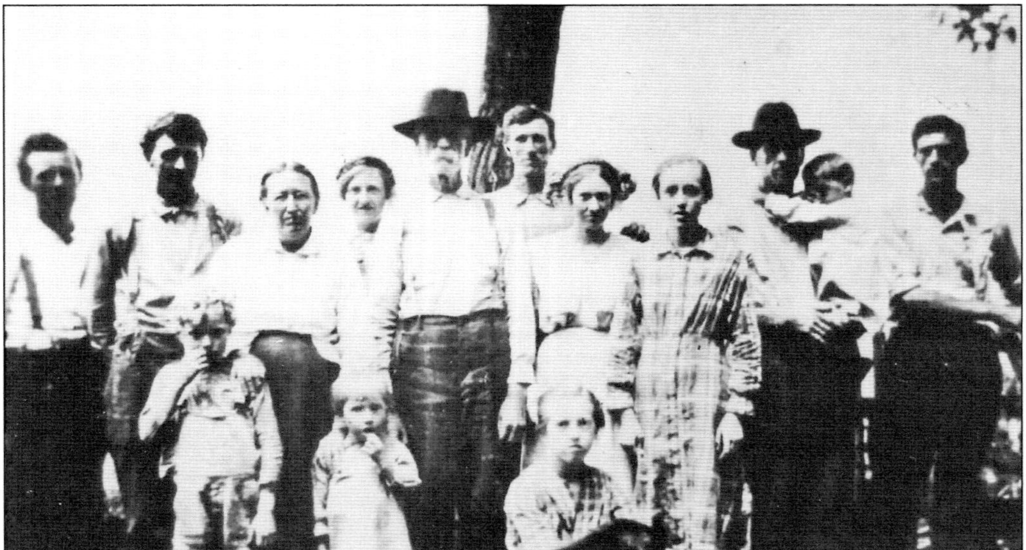

This is the family of William Hurst Dunn and Dorothy C. Snider Dunn, more descendants of Peter Snider. Peter came up from Fort Henry looking for a site to settle. He liked the Tuckaleechee Cove lands and the river valley, but as the Indians were not entirely peaceful, he retreated to come back another day. The house Peter Snider built in Tuckaleechee Cove around 1801 still stands today near the junction to Dry Valley. (Courtesy of Great Smoky Mountains National Park.)

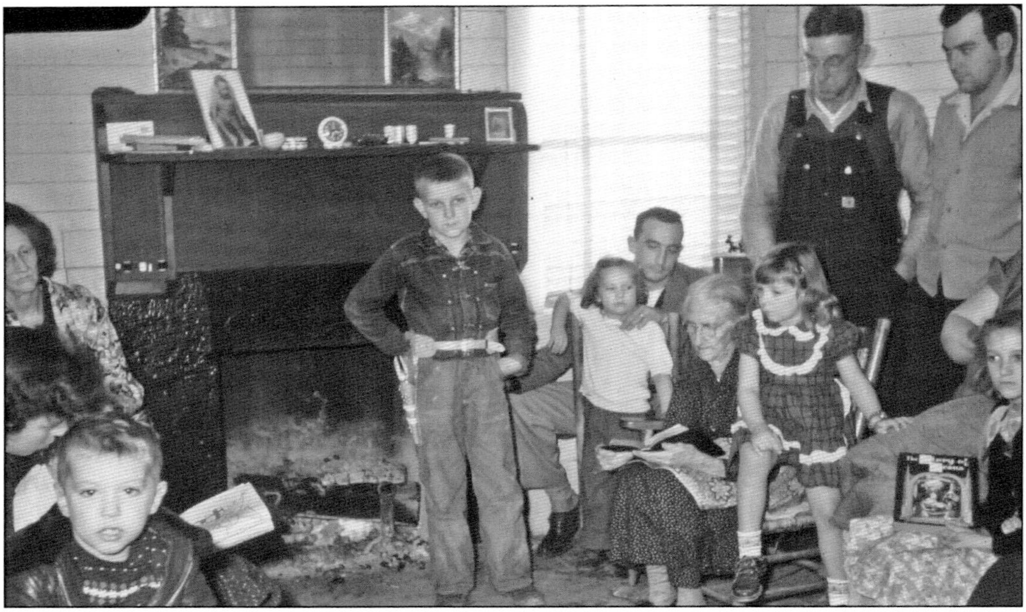

In this image, the McCampbell family is gathered on Christmas Day 1953. This was an important time of year for families to come together for fellowship. Delia Dunn McCampbell has her Bible open and is reading to the family. The McCampbell family resided in the Dry Valley area of Tuckaleechee Cove. (Courtesy of the William Derris Slide Collection, MS.2123. University of Tennessee Libraries, Knoxville, Special Collections.)

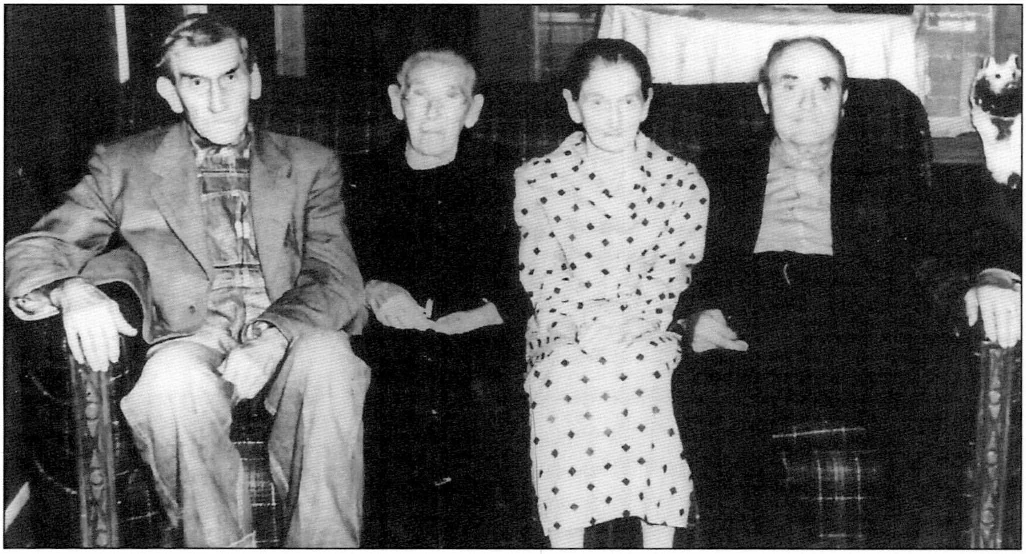

Pictured here from left to right are Jake Dorsey, W. Texanna Burchfield Dorsey, Mary Dorsey Myers, and William A. "Pot Leg" Myers. Pot Leg was born September 10, 1870, married Mary Dorsey on August 20, 1891, and died July 20, 1952. Mary was born on August 6, 1869, and died March 20, 1952. Pot Leg, the son of Daniel H. "Dan" Myers and Elizabeth "Betsy" Ann Webb Myers, ran a store near the entrance to the Great Smoky Mountains National Park. The existence of stores in the area enabled families to purchase necessities without having to travel a long distance. (Courtesy of Paulette Ledbetter.)

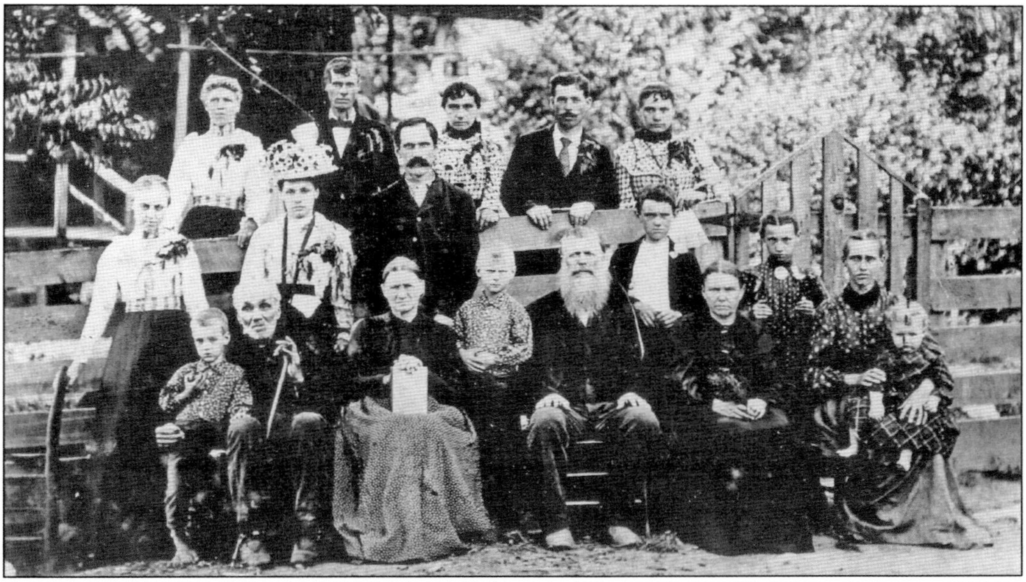

John and Nancy Wear Caylor pose here with their children and grandchildren. As in many older pictures, some of the individuals are holding Bibles. Judging by their dress, this was probably a special occasion. John and Nancy's daughter Mary Josephine married Daniel Edward Headrick, their son Eli Houston married Lula Myers, and their daughters Ella Rosette and Sara Susan married Daniel Alexander Headrick and Clifford Cotter respectively. (Courtesy of Marilyn Myers Byrd Photograph Collection.)

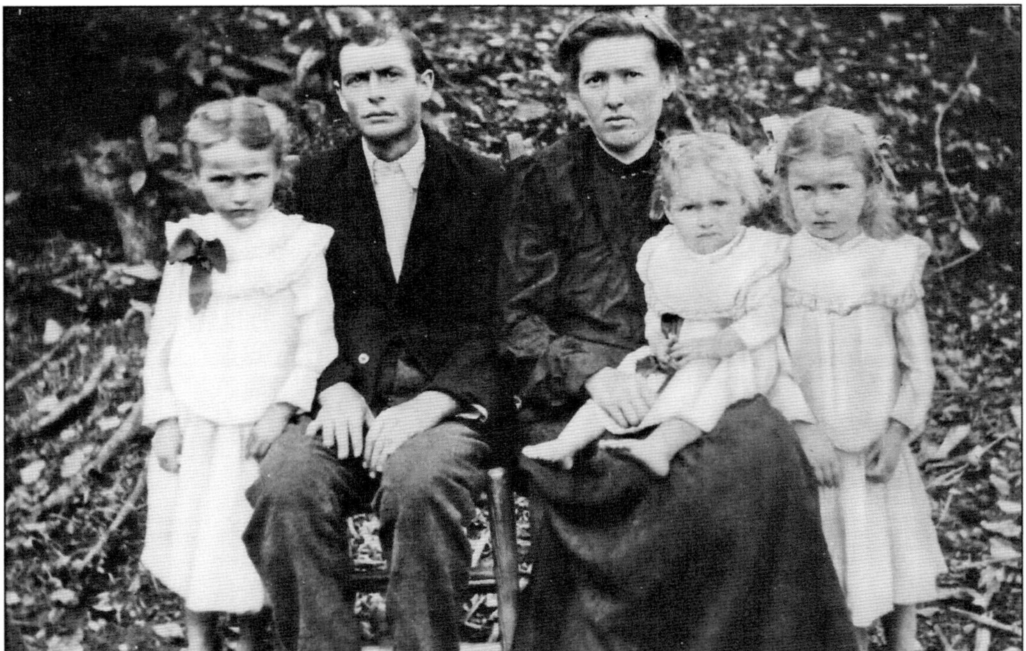

George and Mary Law Caylor are pictured with their children, Angie, Pearl, and Lela. George was born on May 20, 1876, in Tuckaleechee Cove, the son of George Caylor III and Elizabeth Brickey Caylor. He died on March 12, 1971. (Courtesy of Marilyn Myers Byrd Photograph Collection.)

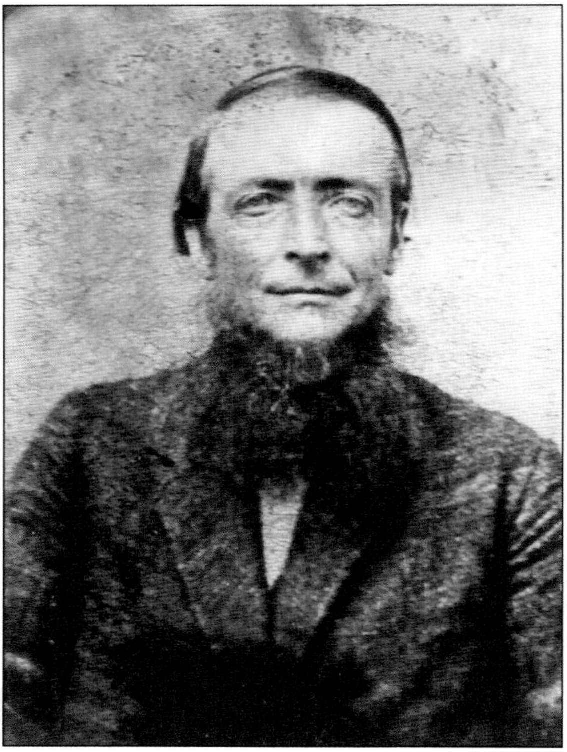

George "Hinkey" Caylor III married Elizabeth Brickey and lived in the Carr's Creek section of Townsend. This was in close proximity to the Lily Barn, now a tourist attraction. (Courtesy of Marilyn Myers Byrd Photograph Collection.)

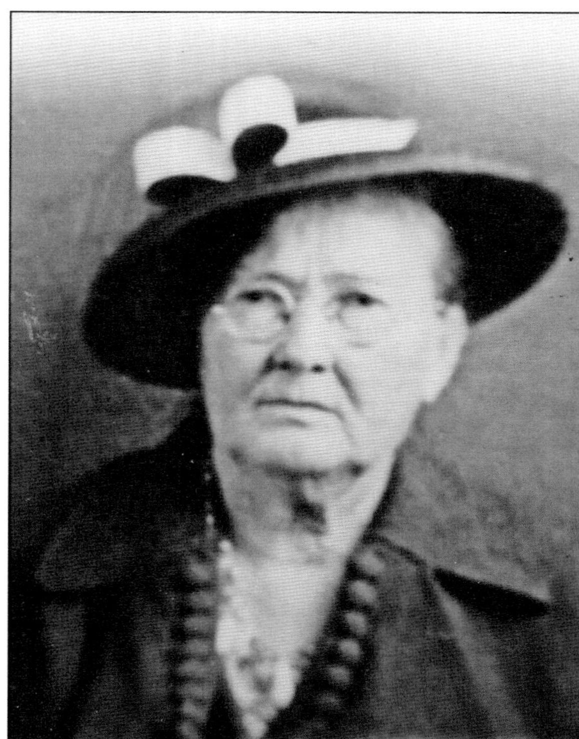

Mary Law Caylor was the daughter of John Wesley Law. She sent her son Lloyd to medical school. Mary sold eggs to grocery stores and hotels; many stores in the community took her eggs on a barter system, trading a good for a much-needed article at the store. (Courtesy of Marilyn Myers Byrd Photograph Collection.)

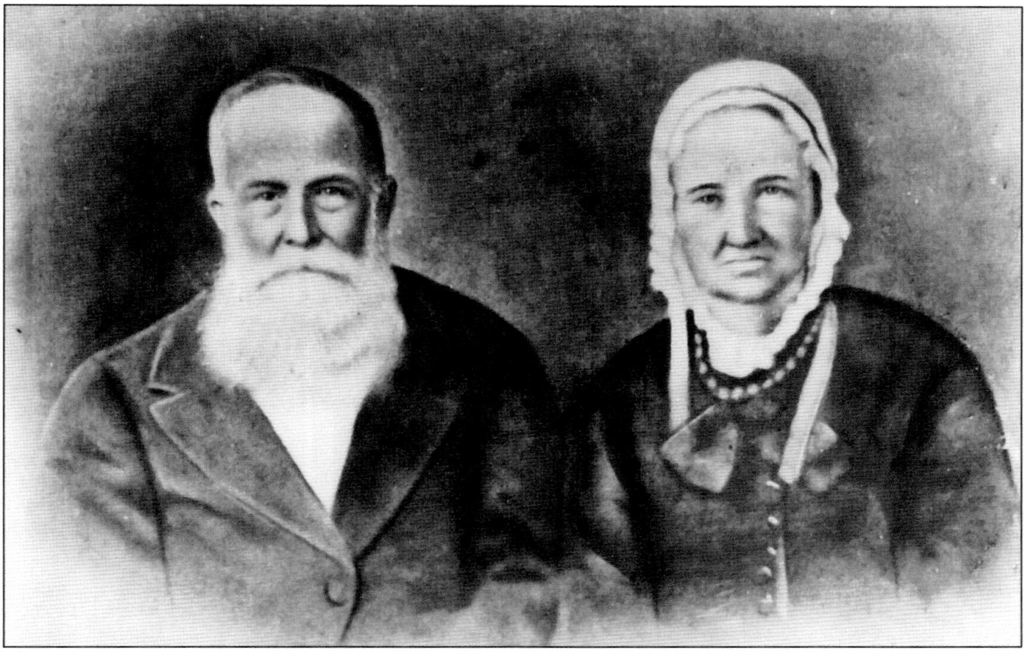

Ailey Caylor and his wife, Susan Thomas Caylor, lived on Carrs Creek on land inherited from his father, George Caylor Sr. Ailey lived in the Tuckaleechee Cove area from 1809 to 1894 and raised eight children. (Courtesy of Marilyn Myers Byrd Photograph Collection.)

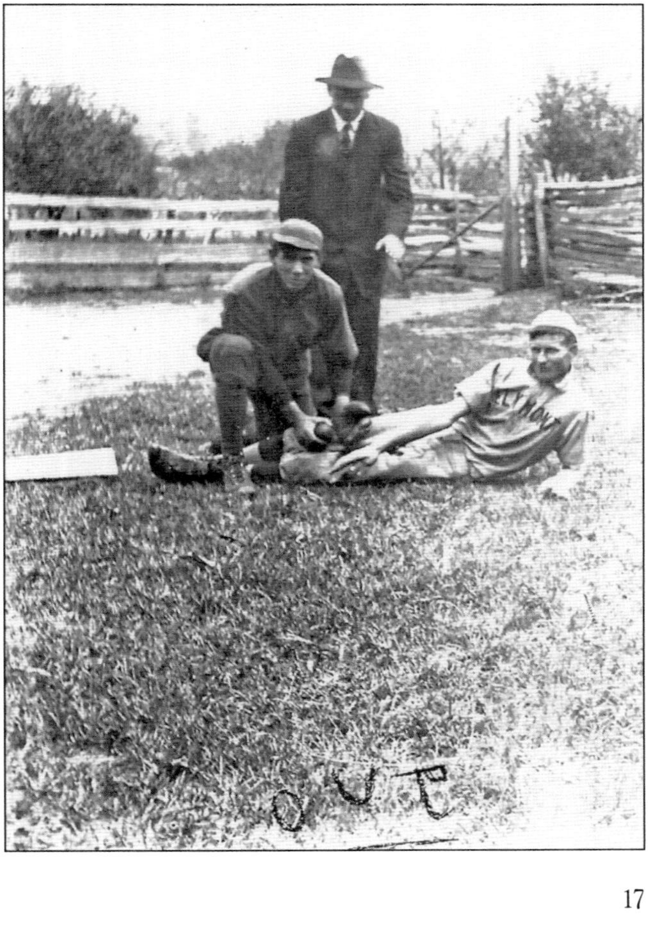

In this image, Bruce Meyers is standing, Bill Myers is kneeling, and Sheridan Wear is "sliding" to home plate. Living in the city of Townsend was not always all work and no play; baseball was a popular game for the men and boys of the town. (Courtesy of McNiell Family Collection.)

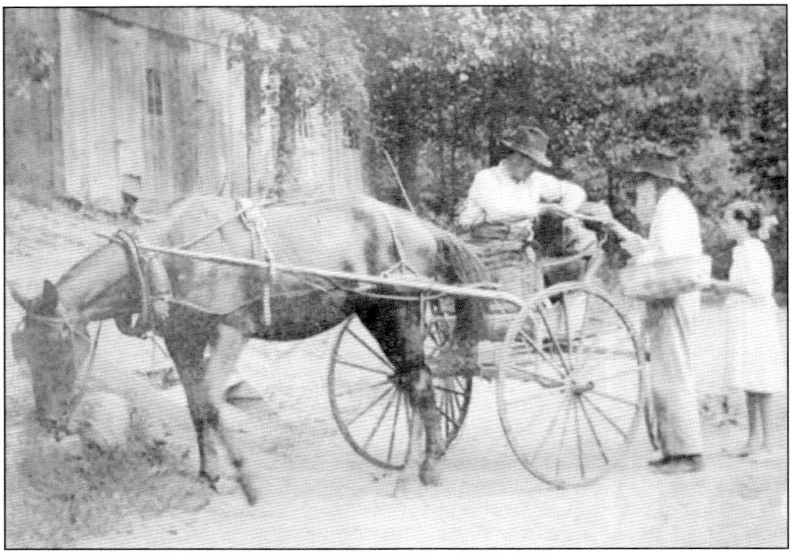

John McCampbell, seen here with his horse and buggy, was employed as the mail carrier until February 12, 1932. In this image, he is delivering mail to E.J. Kinzel, the owner of the Kinzel Springs Hotel, and to Annias Hodge, a nearby neighbor. (Courtesy of Missy Tipton Green.)

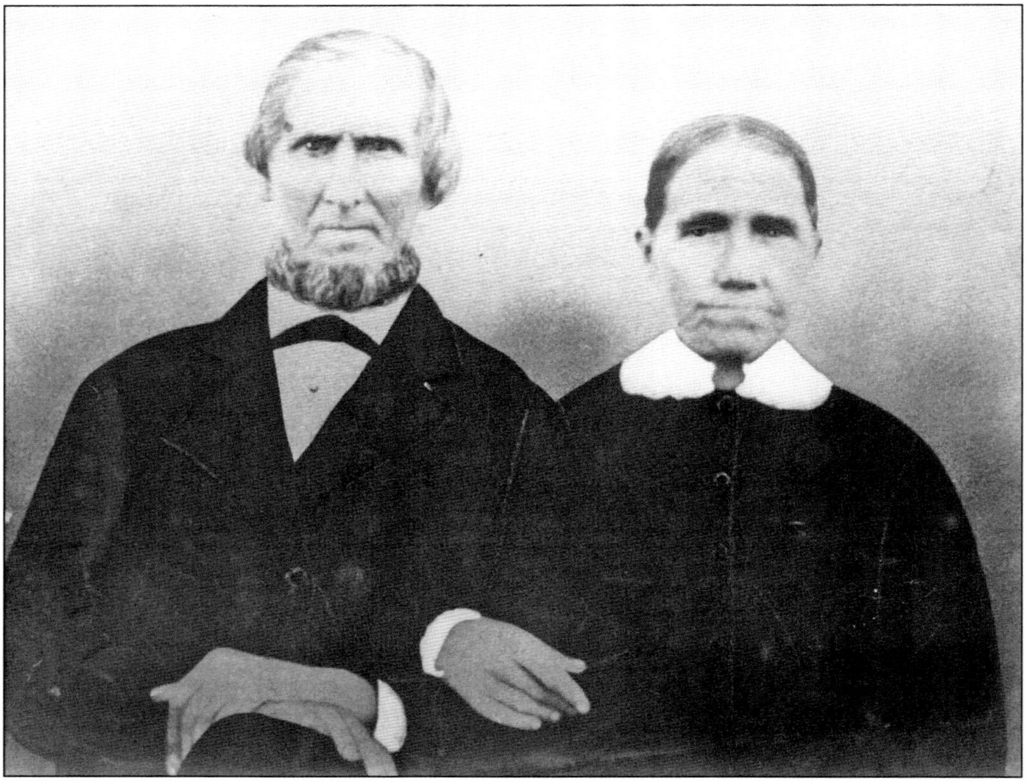

This photograph shows William and Mary Walker Myers. William moved with his parents, John and Jane Dunn Myers, from Greene County, Tennessee, to Tuckaleechee Cove. On January 23, 1847, he married Mary Walker. William had little formal education; however, he was a very shrewd dealer in land, cattle, and other business. He bought many farms in the Tuckaleechee Cove area, and he and Mary raised 11 children. (Courtesy of Marilyn Myers Byrd Photograph Collection.)

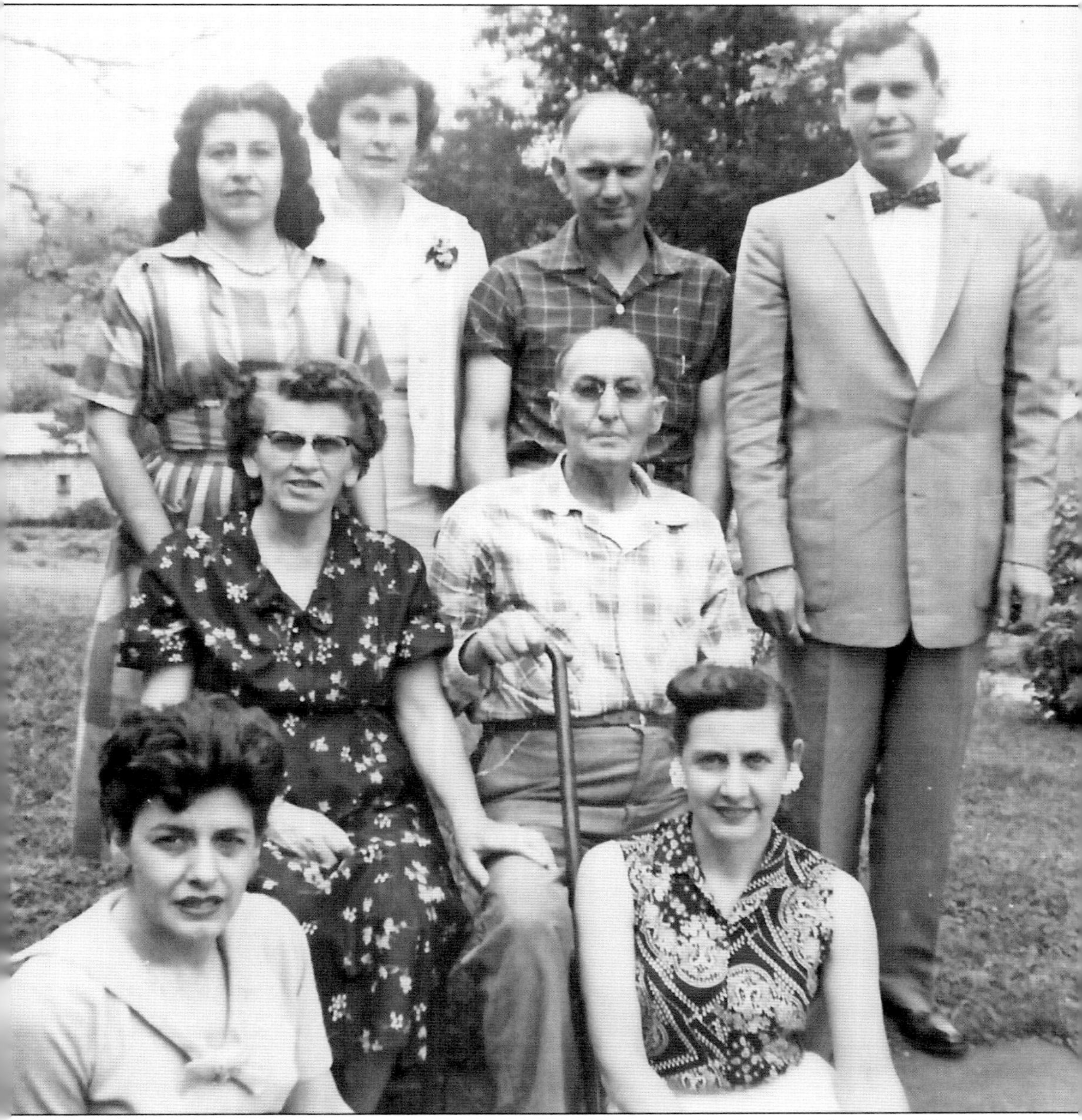

Elder Shade and Iva Sands Tipton are pictured with sons Harold and Robert and daughters Hazel, Mabel, Edna, and Pearl. Elder was born and raised in Cades Cove, as was Iva. He was born on May 5, 1890, to John and Harriett Burchfield Tipton, and he died on November 30, 1958. He and Iva are both buried in the Tuckaleechee Primitive Baptist Church cemetery. After leaving Cades Cove, the Tipton family resided on Cedar Creek Road in Townsend. Pictured from left to right are, (first row) Pearl and Edna Tipton; (second row) Iva Sands Tipton and Elder Shade Tipton; (third row) Mabel, Hazel, Robert, and Harold Tipton. (Courtesy of Louise Tipton Loan, Bertha Proctor Tipton Collection.)

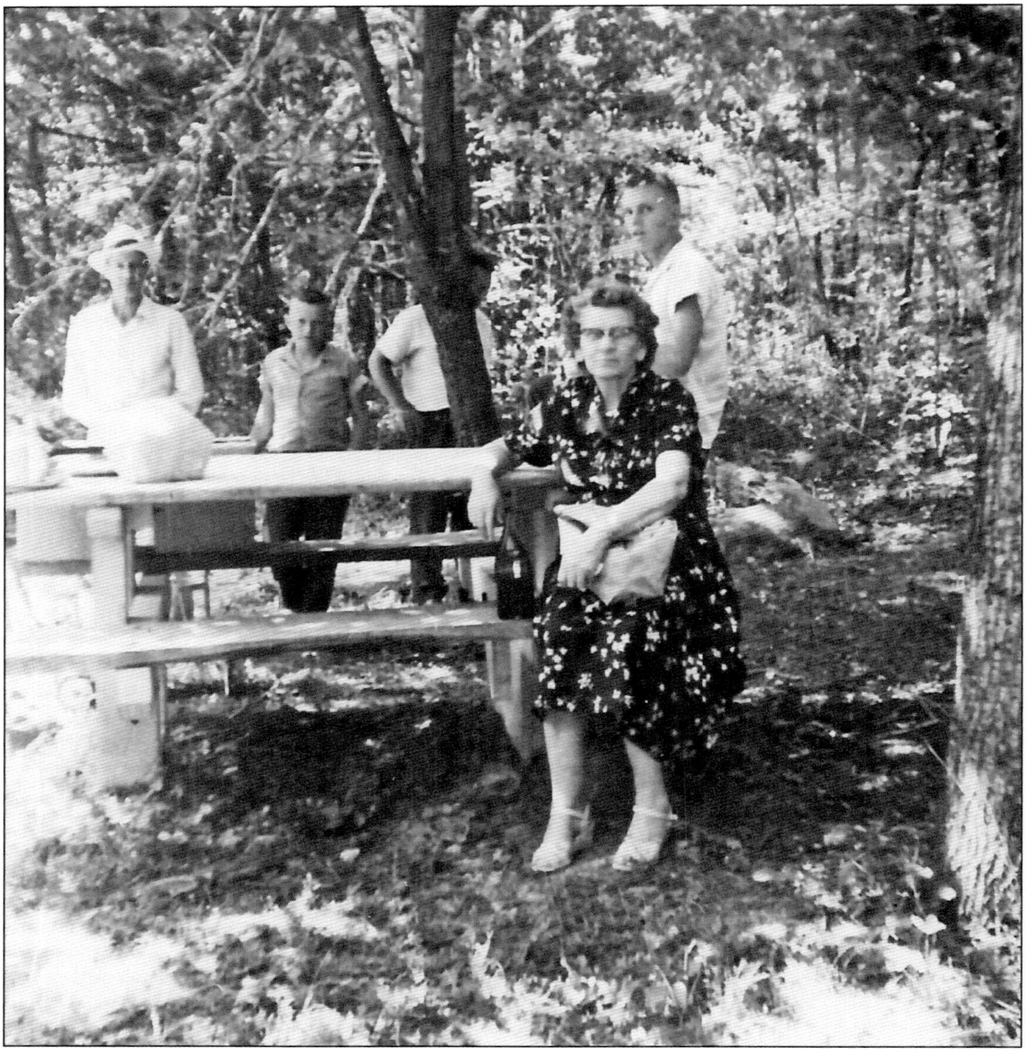

Iva Texanna Sands Tipton was born and raised in Cades Cove, Tennessee; her parents were Henry Anderson and Laura Texanna Burchfield Sands. Iva was married to Elder Shade Tipton. After moving out of the Cades Cove area, the family resided in Townsend. (Courtesy of Louise Tipton Loan, Bertha Proctor Tipton collection.)

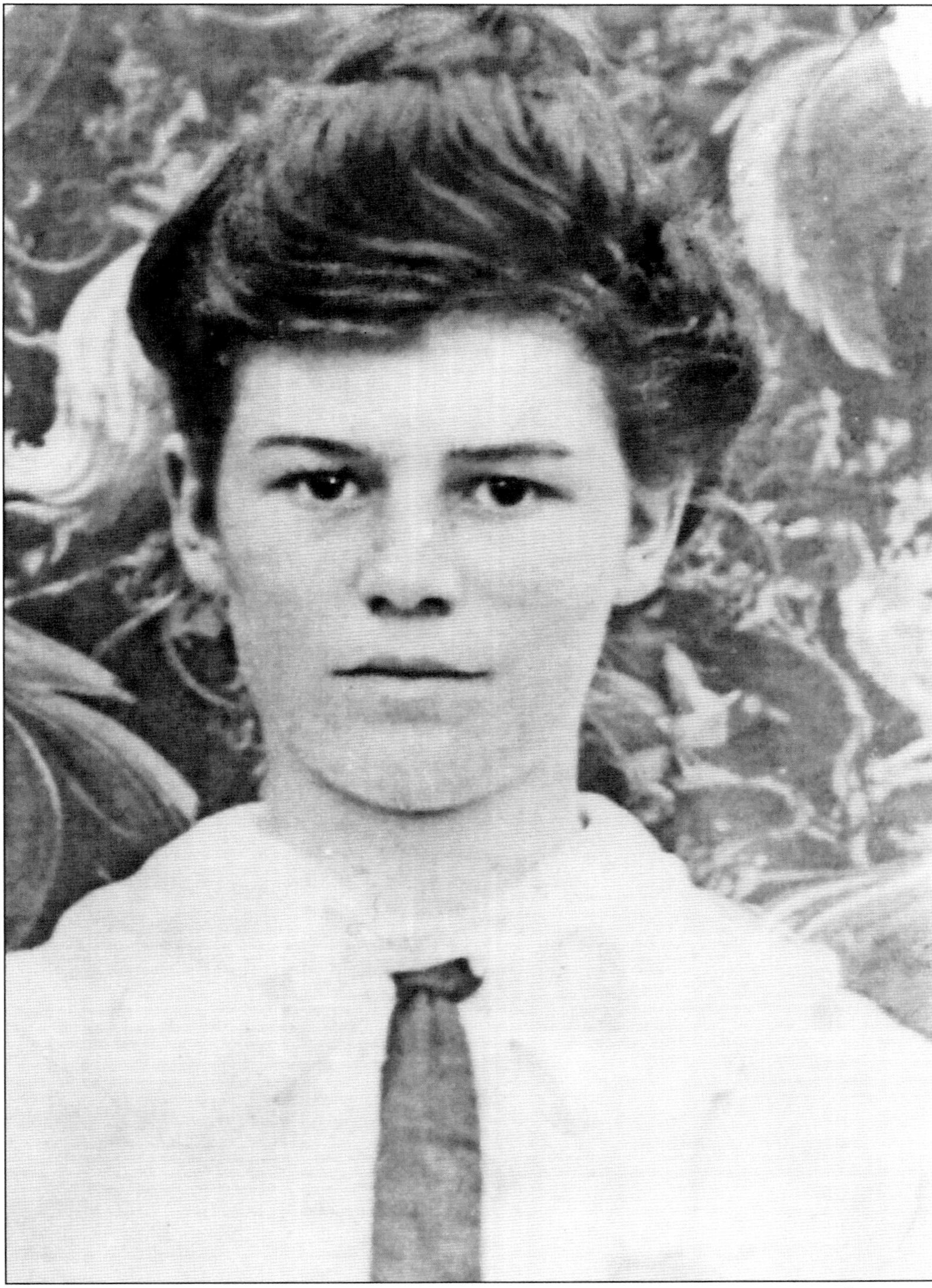

Leona Ellen "Daisy" Morris Effler, daughter of Albert and Cordia Curry Morris, married Harrison Effler on July 19, 1908. Daisy always grew a row or two of cotton to use in the making of her quilts. She gathered goose feathers lying in the yard and saved them to refill pillows. (Courtesy of Paulette Ledbetter.)

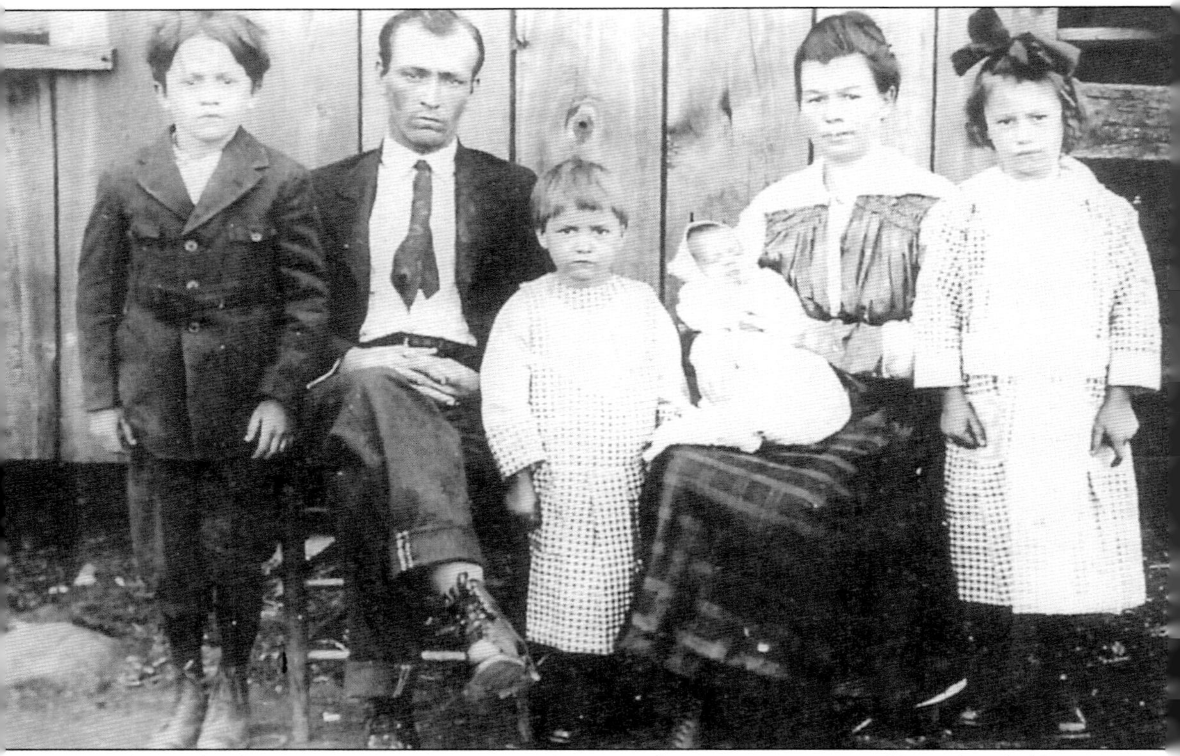

This is the family of Henry Harrison and Leona Ellen "Daisy" Morris Effler. The children are, from left to right, Amos, Zella, Ottis, and Lizzie. Harrison worked a while for Little River Railroad, and the family lived in the area that is now known as Boat Gunnel Road in Dry Valley. (Courtesy of Paulette Ledbetter.)

Daisy Morris Effler and her daughter, Juanita, are seen here. Juanita remembers that her mother always enjoyed decorating graves each year for Decoration Day. Daisy used crepe paper to make flowers in different colors, or if spring flowers were in bloom, she used them to decorate. Juanita's remembers the graves were always so pretty on Decoration Day. (Courtesy of Paulette Ledbetter.)

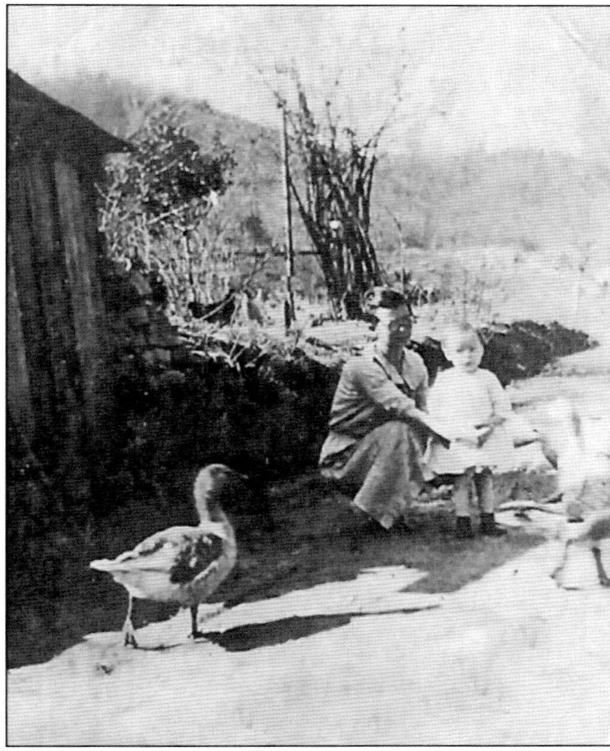

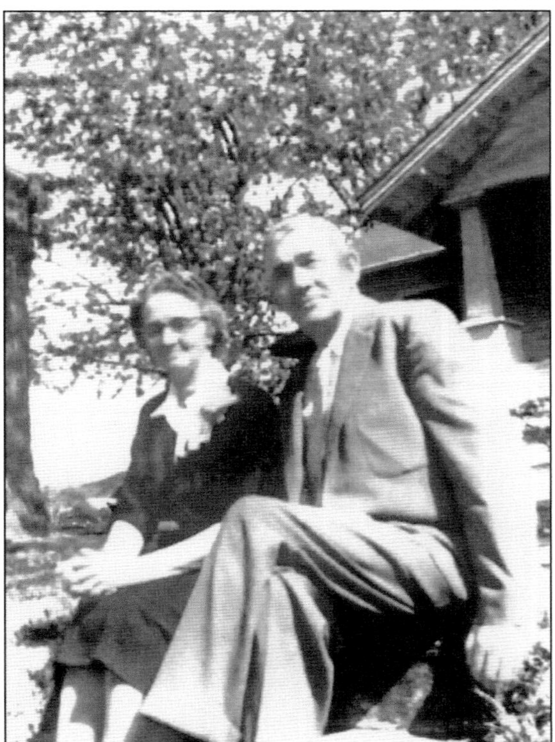

On October 33, 1927, Arville R. Lawson married Mary "Ila" Lowe Lawson. Arville and Ila ran a general store in Townsend. The Lawson house was one of the few homes that had a "real" telephone in Townsend; the phone number was 3. People lined up to use the Lawson telephone during World War II; the call operator in Maryville made a connection and then placed the call. In 1948 or 1949, Arville and Ila bought full interest in the Lawson store. Their son Fred drove a 1.5-ton truck to deliver groceries to customers up in the mountains and coves, carrying 100-pound bags of feed for the cows or 25-pound bags of cornmeal or flour. (Courtesy of Fred Lawson.)

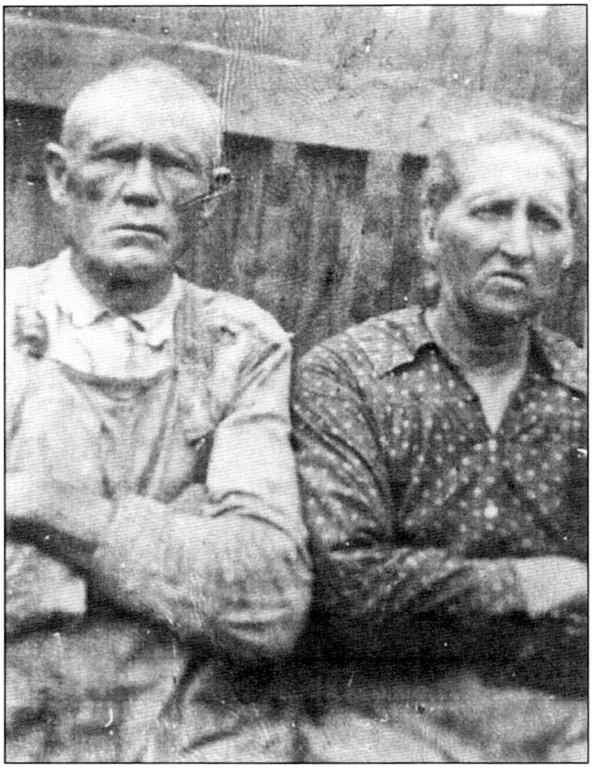

William Henry "Bill" and Mary Caroline ("Cal" or "Callie") Abbott McClanahan raised their family of seven children in Tuckaleechee Cove. Cal was the daughter of Noah and Nancy Hatcher Abbott. Her children with Bill were Sam, David, Maude, Claudia Jane, James Boyd, Mayme, and Ollie. The McClanahan farm was located in the mountainous area of Townsend; there were bear and deer and other animals to hunt, but they were still located in close proximity to Little River Lumber Company. (Courtesy of Eddie McClanahan.)

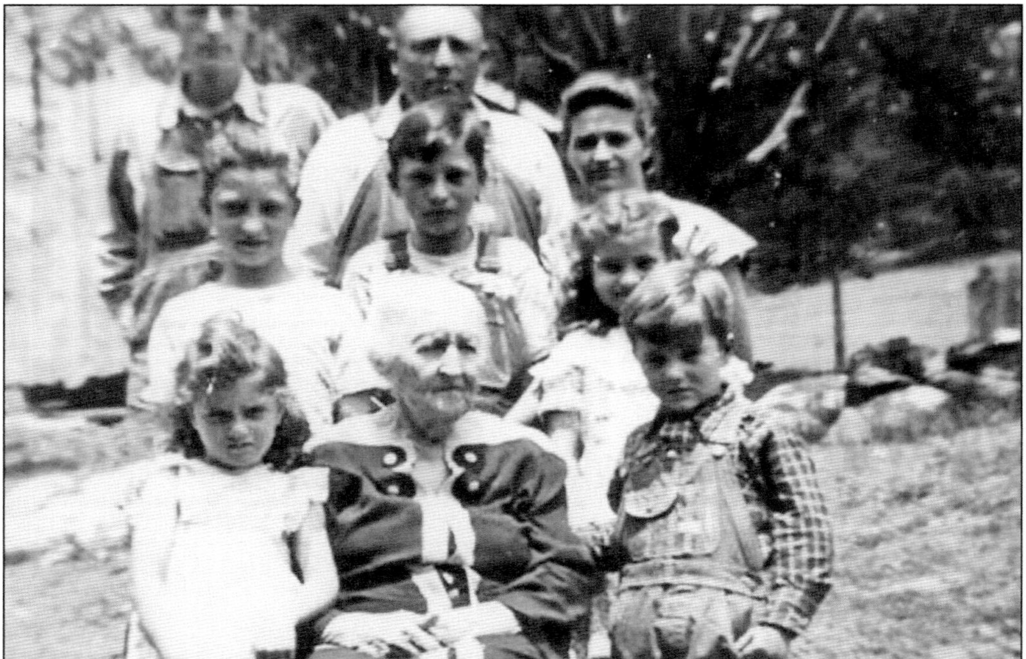

Pictured are Eddie Hodge, James Boyd and his wife, Zella McClanahan, and children John, Billy, Mary, Donna, Mary Cal, and Sam McClanahan. (Courtesy of Eddie McClanahan.)

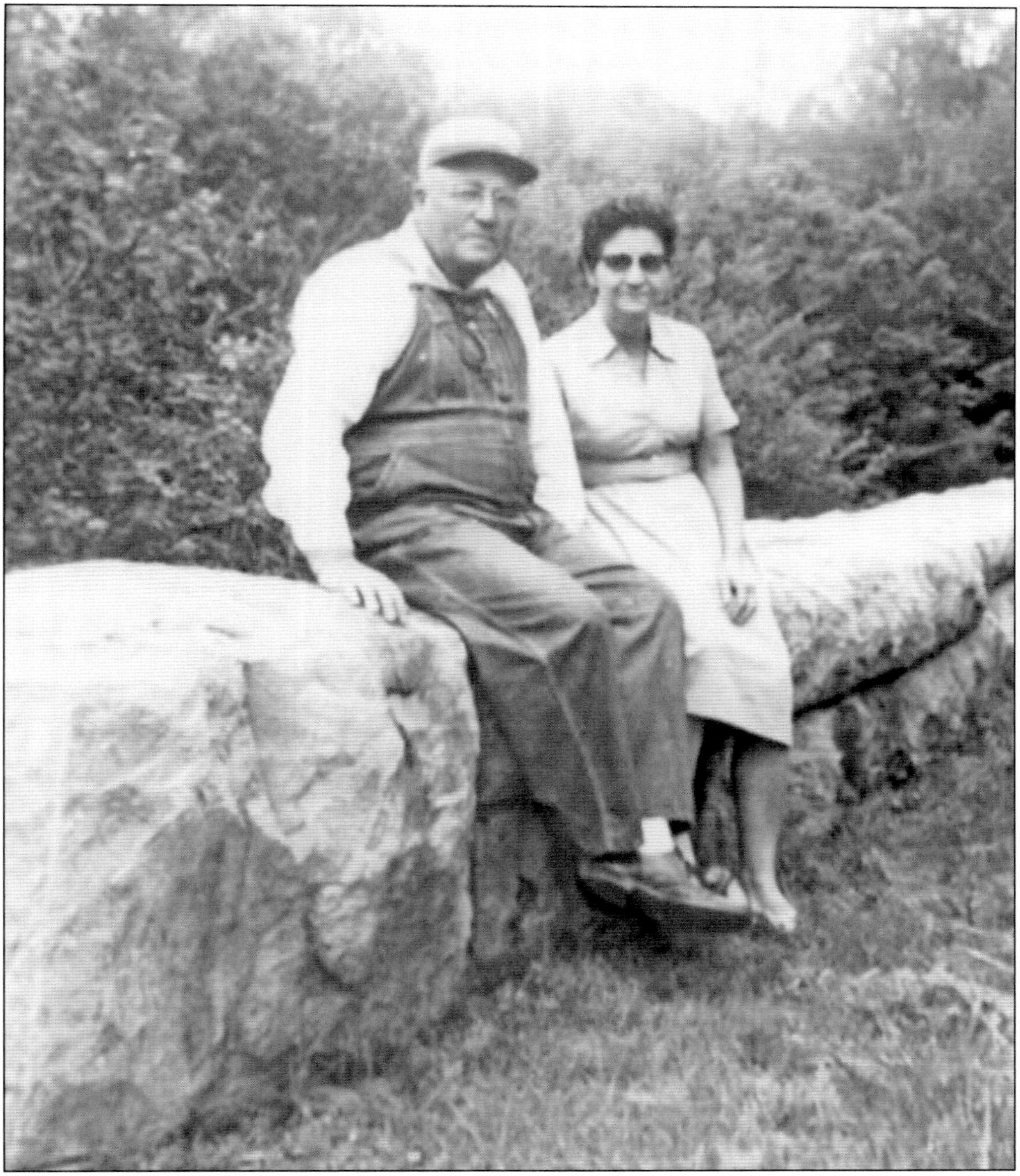

In this photograph are James "Boyd" and Zella Effler McClanahan. Boyd, the son of Billy and Callie Abbott McClanahan, married Zella, the daughter of Harrison and Daisy Morris Effler. (Courtesy of Eddie McClanahan.)

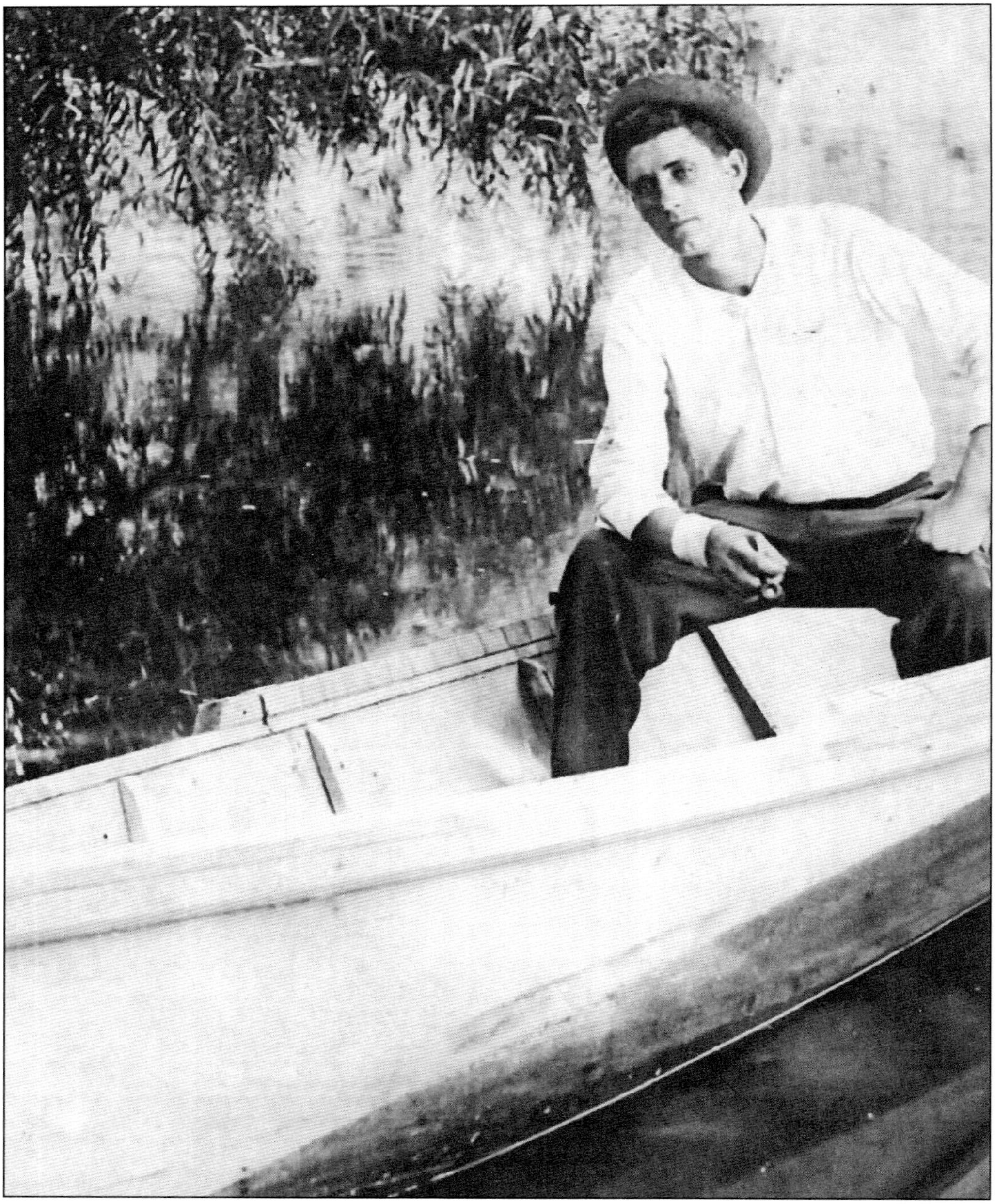

Samuel "Sam" Atchley McClanahan (March 13, 1917–June 26, 1899) was the firstborn of William Henry and Mary Carolina Abbott McClanahan's seven children. At the time of his birth, his family had been in Tuckaleechee Cove for almost 100 years. Sam was just a young boy when the Little River Lumber Company came into the Tuckaleechee Cove area. He grew up around the company and eventually worked for it. (Courtesy of Eddie McClanahan.)

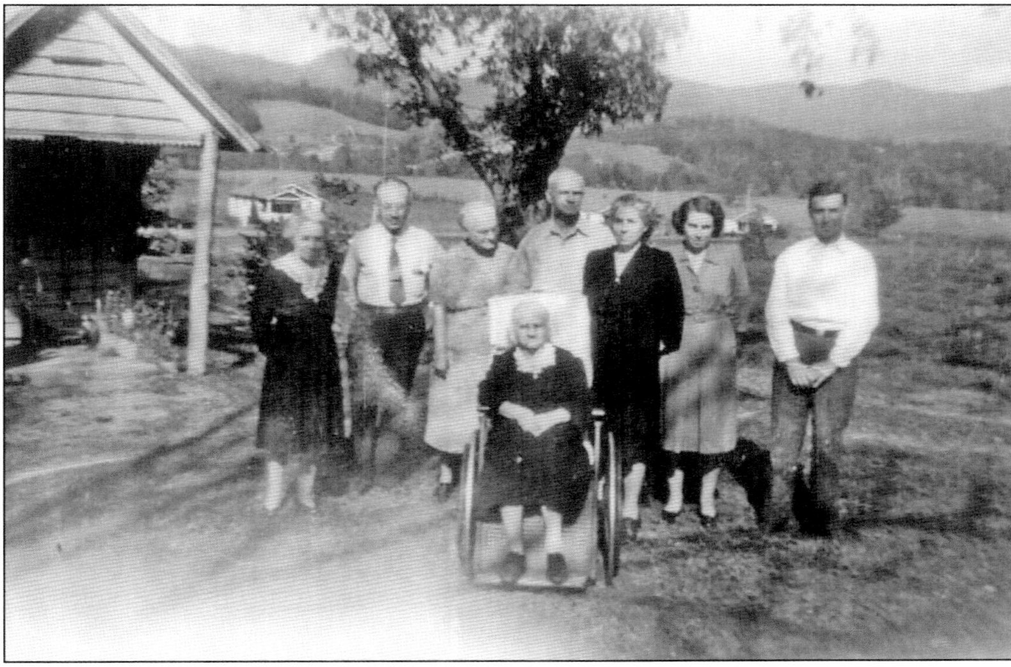

Pictured from left to right are Blanche Myers Webb, Earl Myers, Mary Myers Jones, Bruce Myers, Stella Myers McNiell, Mildred Myers Carnes, and Leonard Myers. They are the children of Joseph and Laura Feezell Myers; Laura is seated in the front center. (Courtesy of Marilyn Myers Byrd Photograph Collection.)

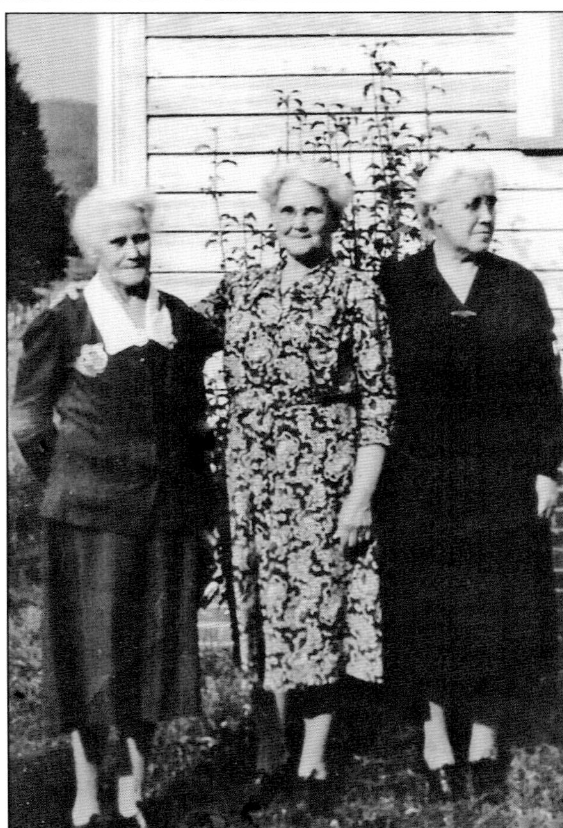

These daughters of John "Sleepy" and Peg Bird Myers are Margaret "Toy" Alfa Myers Dorsey (March 26, 1879–December 7, 1950), Mattie "Matt" Winona Myers Feezell (January 6, 1877–October 21, 1967), and Sara Jacob Jemina "Mina" Farmer (August 28, 1880–January 3, 1956). (Courtesy of McNiell Family Collection.)

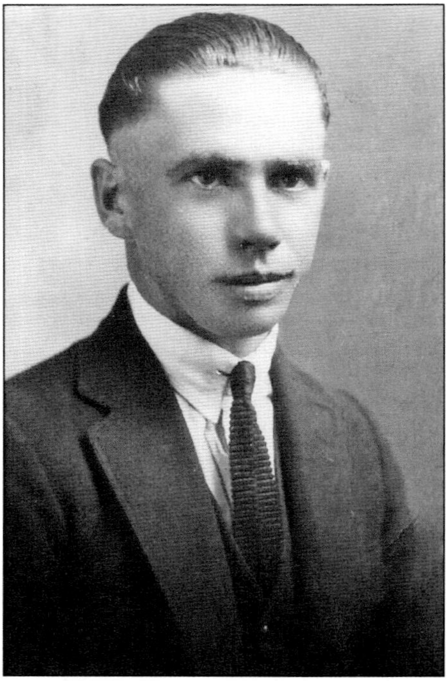

When Lee Myers began working for the Aluminum Company of America (Alcoa), he was assigned as a security guard for transporting company officials to the lodge at Calderwood Dam. (Courtesy of Marilyn Myers Byrd Photograph Collection.)

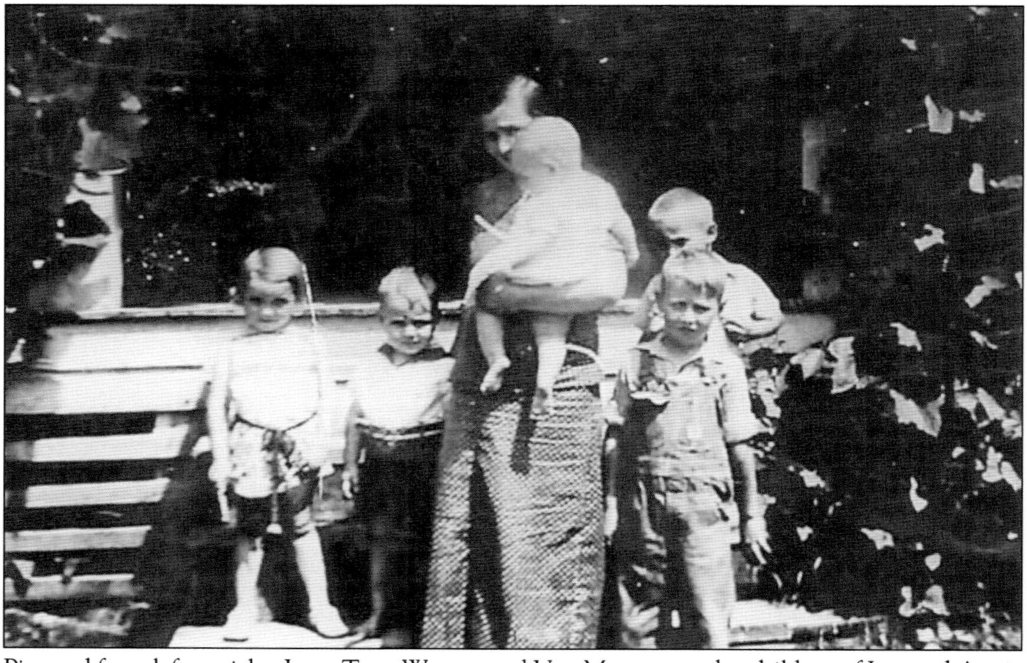

Pictured from left to right, Jean, Tom, Wayne, and Van Myers were the children of Lee and Angie Myers; Angie is pictured holding Marilyn. Lee and Angie's first home in 1930 was one of the Little River Lumber Company's setoff cabins from Elkmont. Shortly thereafter, Lee accepted a job as custodian at Townsend School and rented a small white frame house on Needmore Lane on property adjoining the school. All five of the children were born while living on Needmore Lane; in 1939, the family moved to Myers Lane. (Courtesy of Marilyn Myers Byrd Photograph Collection.)

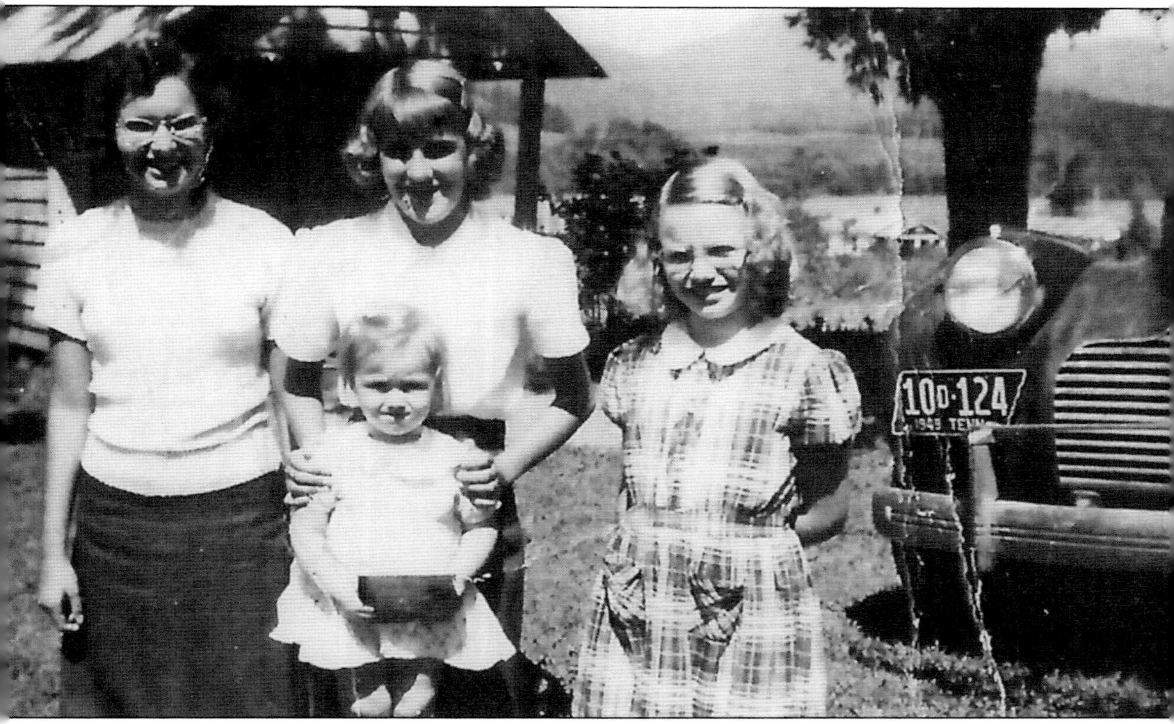

From left to right, Shirley McNiell, Joe Hall, and Marilyn Myers with Donna Porter in front are pictured at the home of Joe and Laura Feezell Myers in about 1950. The young ladies kept busy by jumping rope, shooting marbles, and playing hopscotch and jack rocks. They also had chores to keep busy with during the day. (Courtesy of Marilyn Myers Byrd Photograph Collection.)

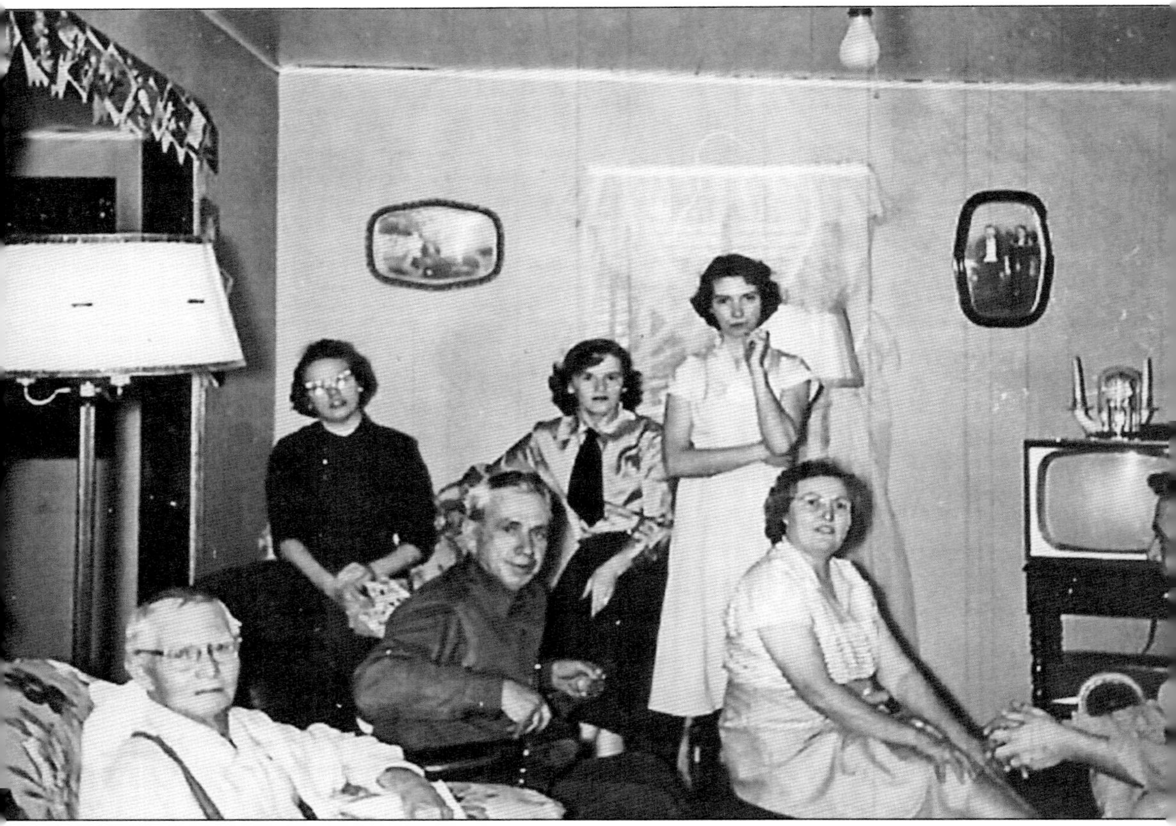

This image captures Christmas at the home of Lee and Angie Myers. In it are, from left to right, George Caylor (Angie's father), Marilyn, Lee, Hilda, Jean, Angie, and Wayne. Christmas was always a time for families to come together for fellowship and good food. Preparations for the Christmas dinner started weeks before the holiday. (Courtesy of Marilyn Myers Byrd Photograph Collection.)

Leonard and Lee Myers enjoyed the livestock on the family farm. After chores were done for the day, there were games and activities such as playing marbles, ball, or cowboys and Indians. (Courtesy of Marilyn Myers Byrd Photograph Collection.)

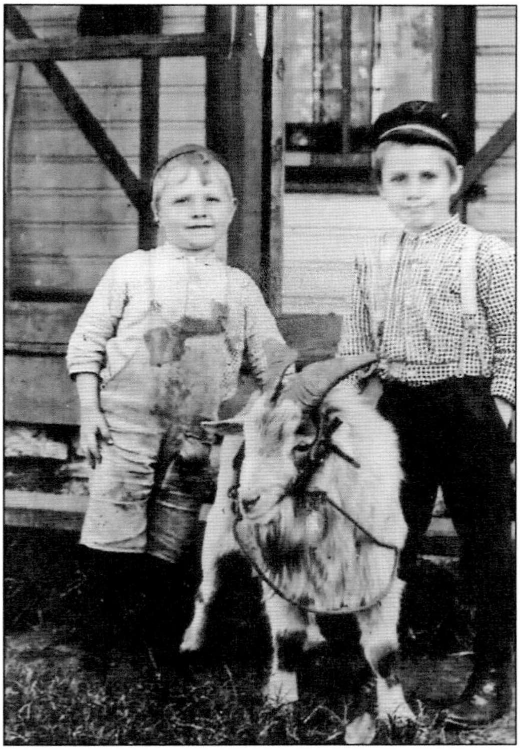

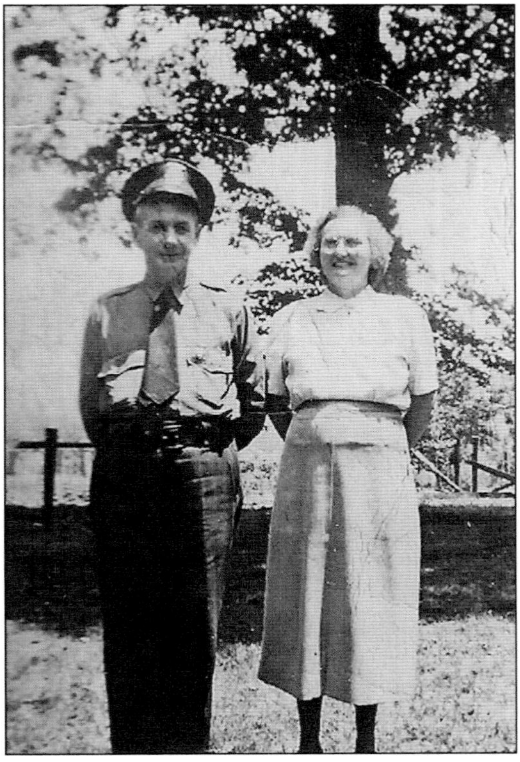

Here are Lee and Angie Caylor Myers. Angie was a schoolteacher at many of the local schools. Lee worked as a guard at Alcoa, first at Calderwood, then at the company's North and South Plants. During the Depression years, he worked as a janitor at Townsend School. In the mid-1930s, he was a local deputy. (Courtesy of Marilyn Myers Byrd Photograph Collection.)

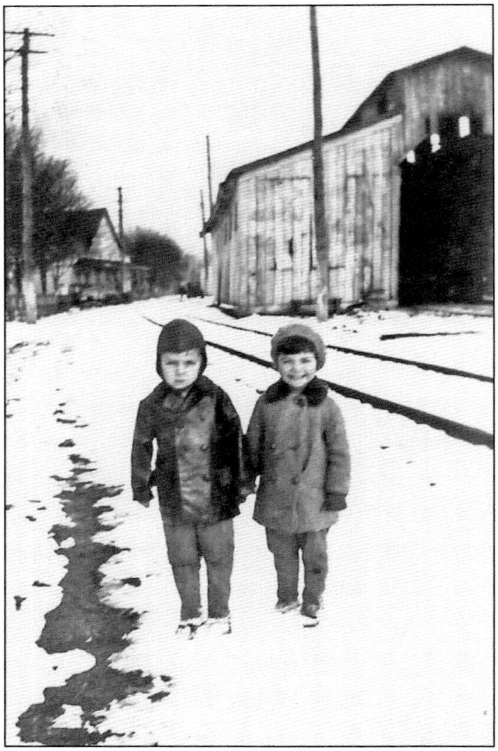

Stuart McNiell Jr. and Bobbie McNiell are photographed in front of the Company Store and the Little River Lumber machine shop. Townsend was a small community, and the company store was a gathering place for all the locals. The McNiell children lived in close proximity to the store and made daily trips there. Pictured below, from left to right, are Stuart McNiell Sr. with Joe, Bobbie, and Stuart Jr. at the machine shop in 1931. (Both, courtesy of McNiell Family Collection.)

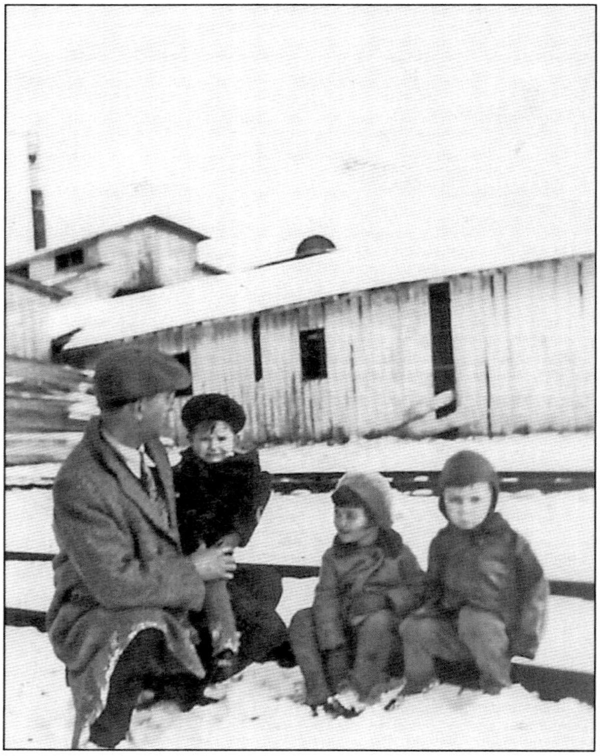

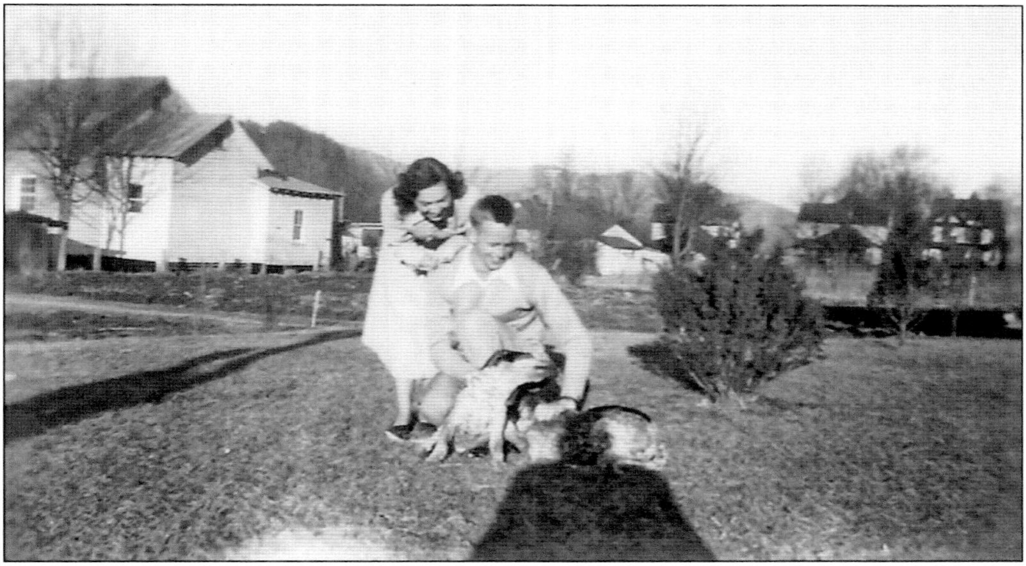

George "Sug" and Barbara "Bobbie" Jean McNiell Handley were married on June 2, 1951. Barbara was born on May 13, 1927, in Townsend to Stuart Sr. and Stella Myers McNiell. Her father was the Little River Lumber Company bookkeeper and superintendent. (Courtesy of McNiell Family Collection.)

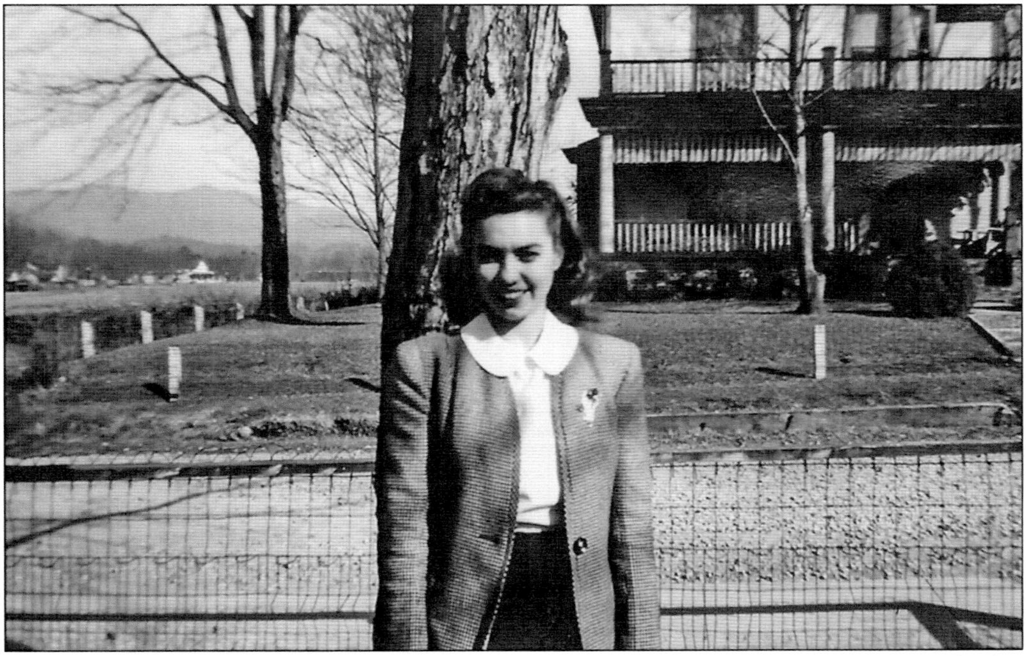

Bobbie McNiell Handley stands in front of the George Townsend house on Needmore Road. This was one of several stately homes built for officers in the Little River Railroad and Lumber Company. It was moved next door to the former Margaret Townsend Church when the new highway was built in the 1950s, and this stately home became the residence of Homer Bradshaw. A small theater once existed where the Bradshaw home now stands. In the distance are the Townsend Mercantile Store and a row of homes. (Courtesy of Marilyn Myers Byrd Photograph Collection.)

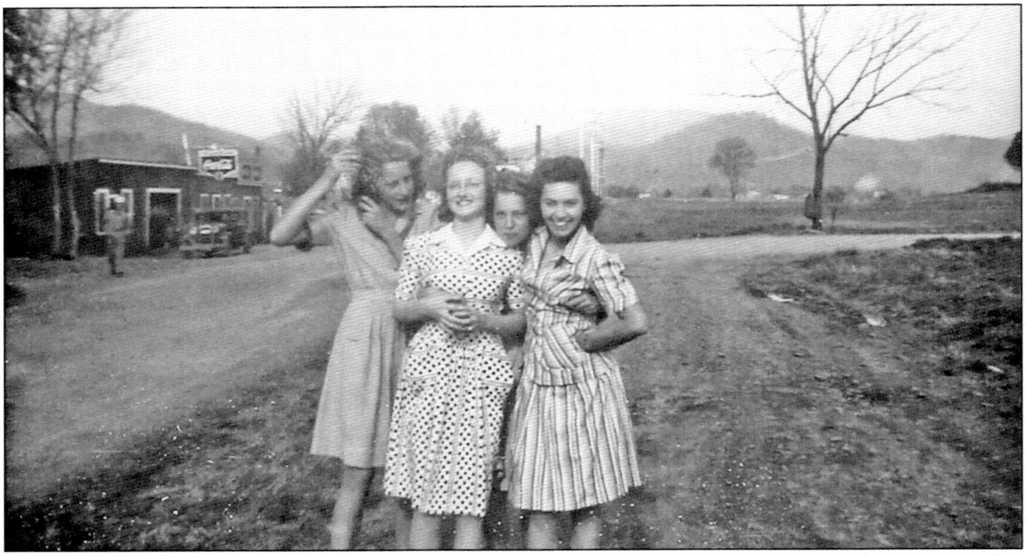

From left to right, Wilma Handley Douglas, Edna Ann "Teen" Myers Greaser, Evelyn Handley Gillespie, and Bobbie McNiell Handley pose in front of Jake Walker's garage and Bill Myers grocery store on the banks of Little River. For several years, Jake Walker was the only mechanic in the community. Though deaf, he was very good at lip reading and understanding hand signals. Bill Myers ran the store next to Walker's Garage and offered good service to the many customers that came by daily. (Courtesy of Marilyn Myers Byrd Photograph Collection.)

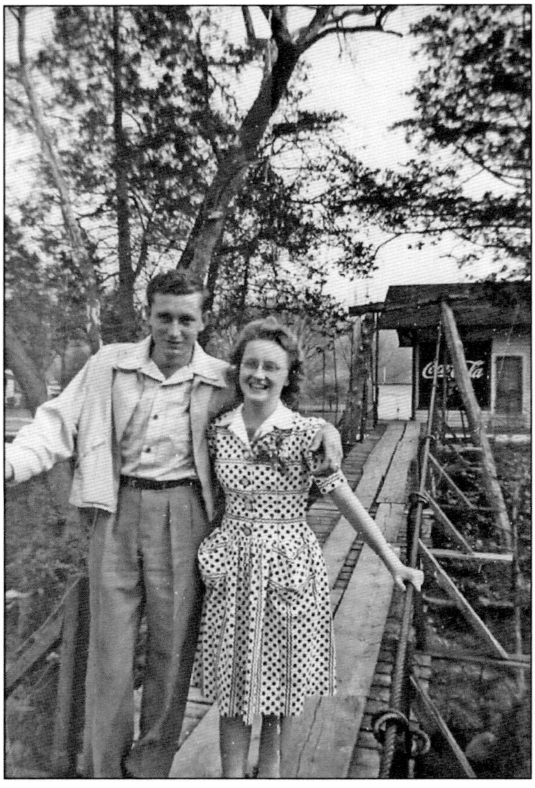

Ralph Greaser and Edna Ann "Teen" Myers stand on a swinging bridge near the Company Store (they later married). Ann is the daughter of Bruce and Louise Myers. Behind the couple is the Little River Railroad depot. Part of it was converted to a sandwich and ice-cream shop operated by the McNiells in the early 1940s. The other part was used by Kate Dykes Dunn as a beauty shop. On Sunday afternoons, all of the teenagers would meet at the swinging bridge. It was also a favorite courting place where boys and girls would meet at a specified time. In later years, Bill Anderson sold groceries and gas from the building. In the early 1950s, part of the structure was converted to a drugstore and soda fountain owned by Paul Ferguson. (Courtesy of McNiell Family Collection.)

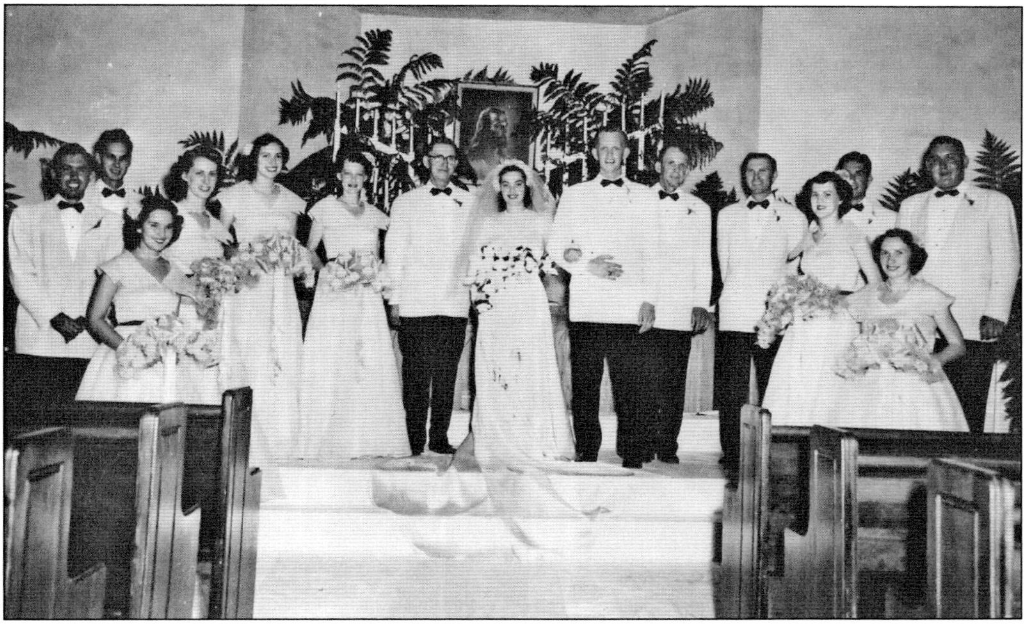

George "Sug" Handley, son of Sam and Leona Gregory Handley, married Barbara Jeanne McNiell, daughter of Stuart P. Sr. and Stella Myers McNiell, on June 2, 1951, at Tuckaleechee United Methodist Church in Townsend. The wedding party included, from left to right, Jane McMillian Baird, Stuart McNiell Jr., Walter Douglas, Wilma Handley Douglas, Shirley McNiell Heatherly, Evelyn Handley Gillespie, Stuart McNiell Sr., Barbara Jeanne McNiell, George "Sug" Handley, Sam Handley, Howard McNiell, Mary Ruth Lawson, Joe McNiell, Helen Myers Bradshaw and Jim Gillespie in the back row. (Courtesy of McNiell Family Collection.)

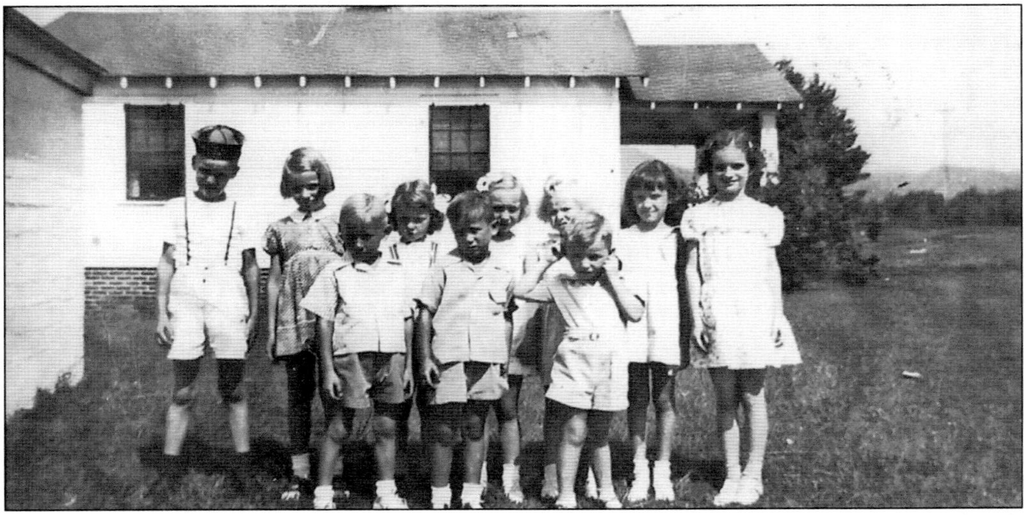

This birthday party for Sandra Vinson took place at the home of grandparents Roy "Bonehead" and Florence Vinson Myers, who lived on the Old Highway 73 in Townsend. Sandra and her parents were visiting from California in about 1945. Pictured are, from left to right, (first row) Lamar Dunn, Jimmy Gregory, and Joe Myers; (second row) Sidney Myers, Marilyn Myers Byrd, Patsy Myers Tipton, Sue Gregory, Sandra Vinson, Betty Farmer, and Juanita Hendrix. (Courtesy of Marilyn Myers Byrd Photograph Collection.)

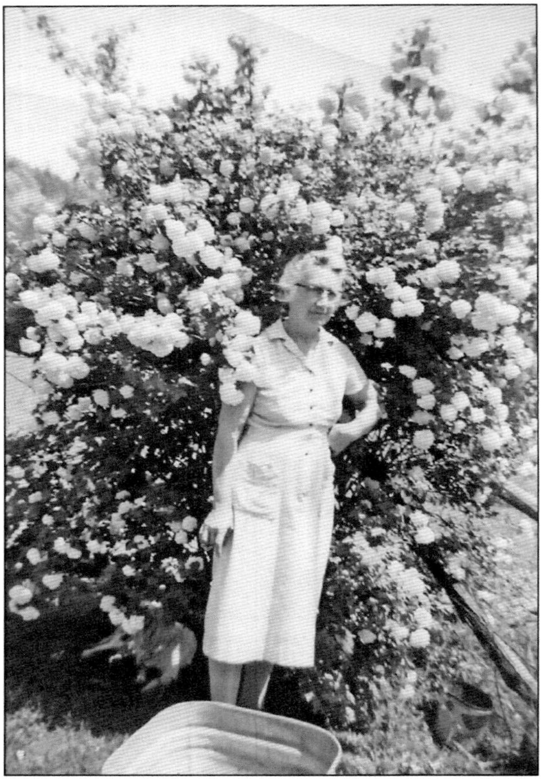

Viola Burchfield Myers, seen here in front of her snowball bush, was born in Cades Cove, Tennessee, on February 22, 1901, to Margaret Tennessee "Tennie" Tipton Burchfield and John H. Burchfield. On October 1, 1920, she married Golman Myers, son of Peter and Margaret Shields Myers. The Myers family moved out of Cades Cove in the early 1940s to a house on Cedar Creek Road in Townsend. Viola enjoyed raising her snowball bushes and other flowers. (Courtesy of Bernard Myers.)

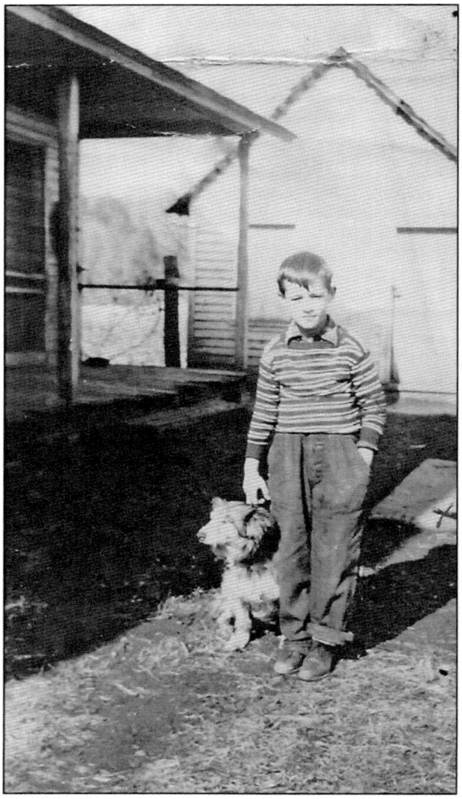

Bernard Myers stands beside his faithful dog "Old Rippy." The location of this picture is the Golman Myers house on Cedar Creek Road in Townsend. Bernard was born in 1936 in Cades Cove, Tennessee, and started his school years there at the Cable School. When he was about eight years old, his family moved to Townsend, where Bernard continued his education at Townsend School. Bernard married Joann Adams. (Courtesy of Bernard Myers.)

Thee, Wayne, and Clee Myers are the children of Golman and Viola Burchfield Myers. They were all born in Cades Cove and all served in the military. The Myers family is known for their good farming abilities. (Courtesy of Bernard Myers.)

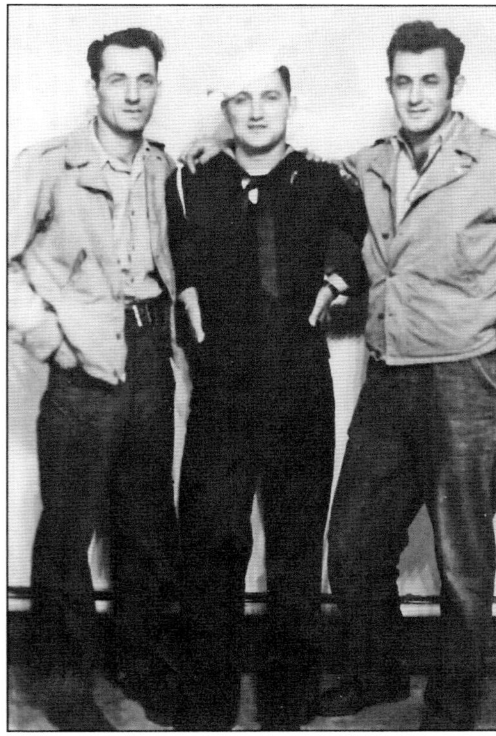

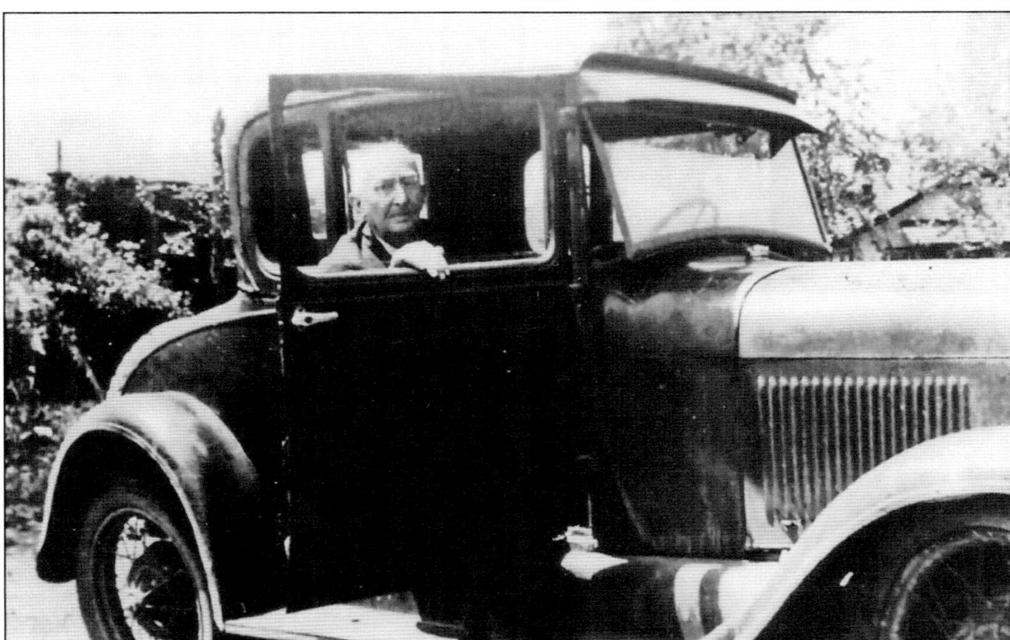

Dr. E.W. Griffin treated many people in the Tuckaleechee Cove and Townsend area. His home office was close to the Little River Lumber Company Store. He charged patients $2; this covered everything from treating pneumonia to delivering babies. His trademark was his flattop straw hat. (Courtesy of Marilyn Myers Byrd Photograph Collection.)

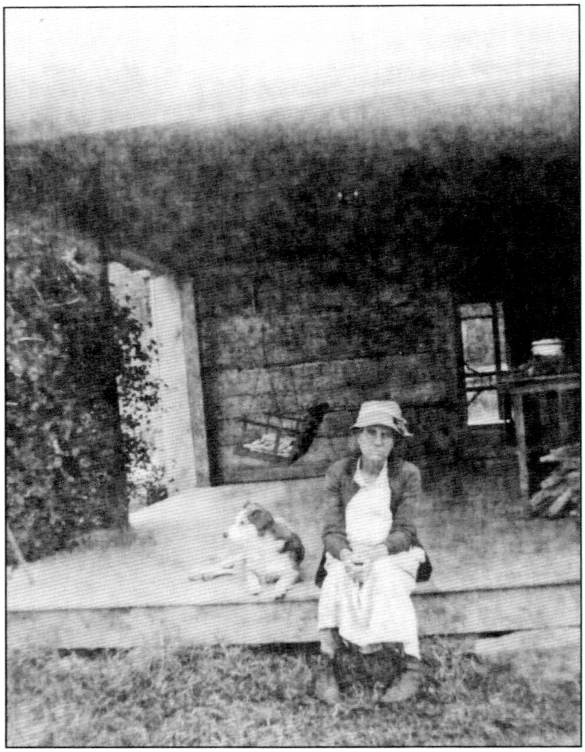

Mary Jane Whitehead Burchfield was born in Cades Cove to Henry and Sara Margaret Boring Whitehead. She married Russell Burchfield also of Cades Cove, and the couple operated a store there for many years. Like many others, they moved from the area to Townsend around 1936 when the Great Smoky Mountains became a national park. (Courtesy of Stephen Webber, John W. Oliver collection.)

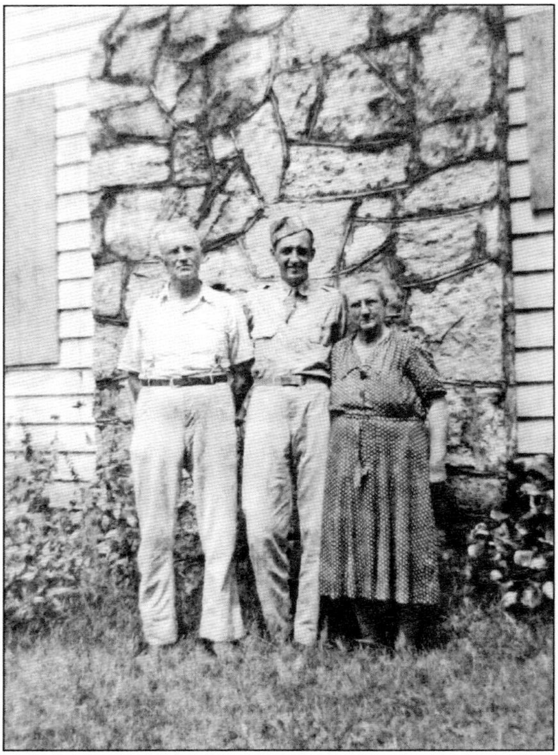

John W. and Nancy Ann Whitehead Oliver are pictured here in Townsend with Clay Oliver (center). Both John and Nancy were born and raised in Cades Cove, which they moved from in 1936. John fought hard to keep his beloved home in Cades Cove from being bought by the state to use for the national park, but he lost his battle and built a house on Bethel Church Road in Townsend. (Courtesy of Stephen Webber, John W. Oliver Collection.)

This photograph shows Otis Tipton and Jane Burchfield. Otis Tipton was born to George and Lula Burchfield Tipton in Cades Cove. Lula died, and George was not able to take care of Otis, so he moved in with his mother's brother Russell Burchfield and his wife, Jane. He continued to live with Jane after the death of Russell; they moved out of Cades Cove in 1936 into a house on Bethel Church Road close to her sister and brother-in-law, Nancy Ann Whitehead Oliver and John W. Oliver. (Courtesy of Stephen Webber, John W. Oliver collection.)

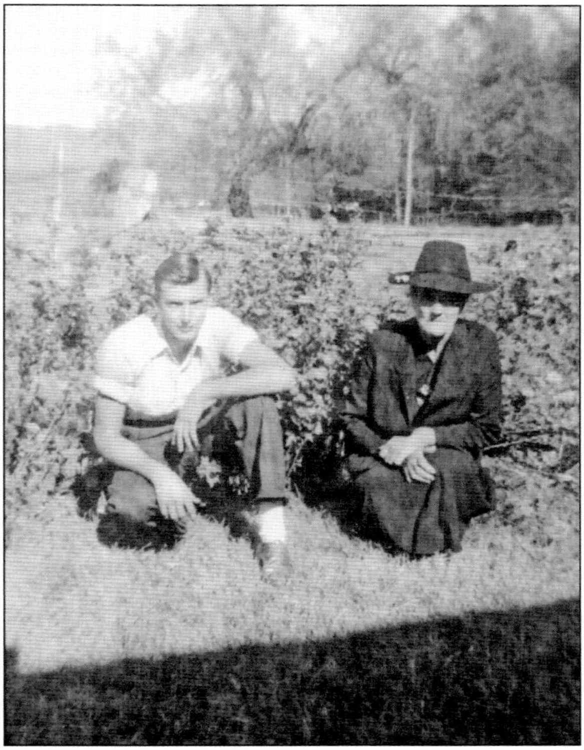

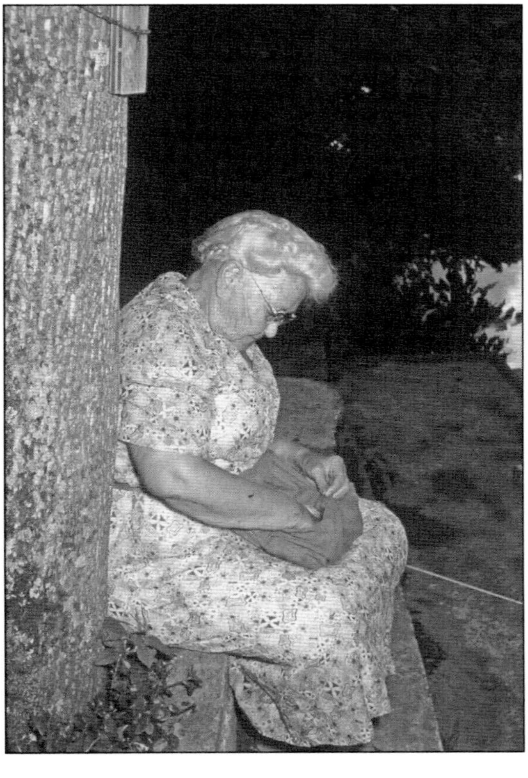

Mattie "Matt" Winona Myers was born on January 6, 1877, to John "Sleepy" and Margaret "Peg" Bird Myers. Matt first married George Emert; after his death, she married George Feezell, son of William and Modena McKeldry Feezell. In this image, she is busy breaking beans, taking them from the pocket of her apron. Aprons were a necessity for women, whether used for gathering eggs, picking vegetables, or to keep their clothing from getting soiled during cooking or cleaning. (Courtesy of the William Derris Slide Collection, MS.2123. University of Tennessee Libraries, Knoxville, Special Collections.)

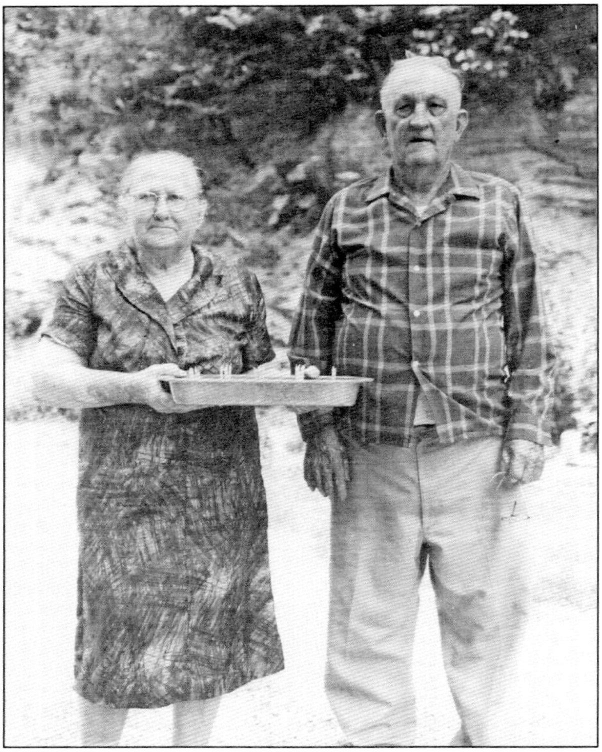

When "Barefoot" Jerry Osborne Effler proposed to Susie "Della" Ledbetter, she had to "study" on his proposal for a while. When she finally came to a decision, she reported to Jerry, "I've decided to take you up on that offer." Jerry's response was, "What offer?" They were married on December 18, 1909, and enjoyed 66 years of marriage before Della passed away on January 18, 1975. (Courtesy of Paulette Ledbetter.)

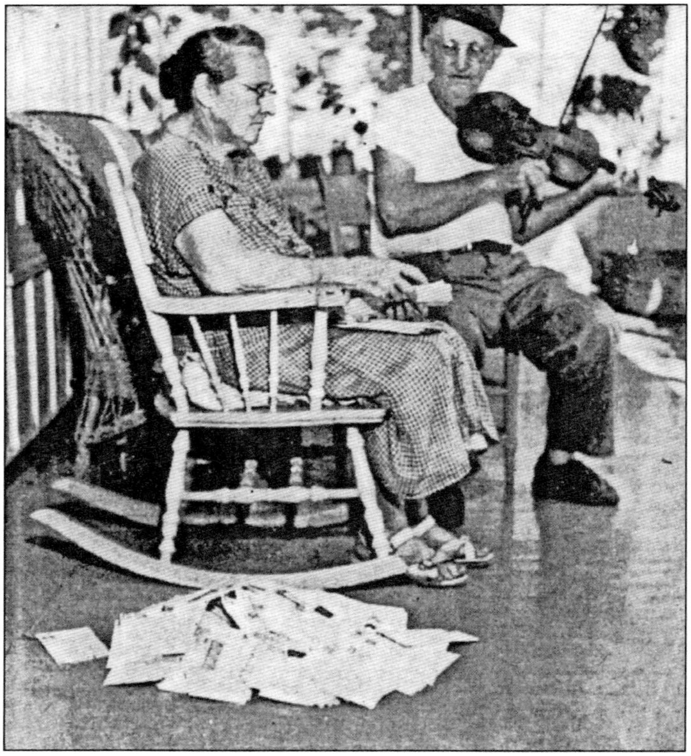

Pictured in 1962, Della and Jerry Effler sit with loads of letters sent to Della asking for her homemade remedies. Della used her mountain medicine for herself and her family, and others heard about it and requested her advice. She did not offer to sell the remedies, but if someone asked, she would give them some good tips. (Courtesy of the *Maryville Alcoa Times*.)

Gibson Tipton was born in the Cades Cove area around 1898 to John Franklin and Harriett Burchfield Tipton. On November 11, 1921, he married Bertha Proctor, also of Cades Cove. Five of their six children were born while living there. (Courtesy of Louise Tipton Loan, Bertha Proctor Tipton Collection.)

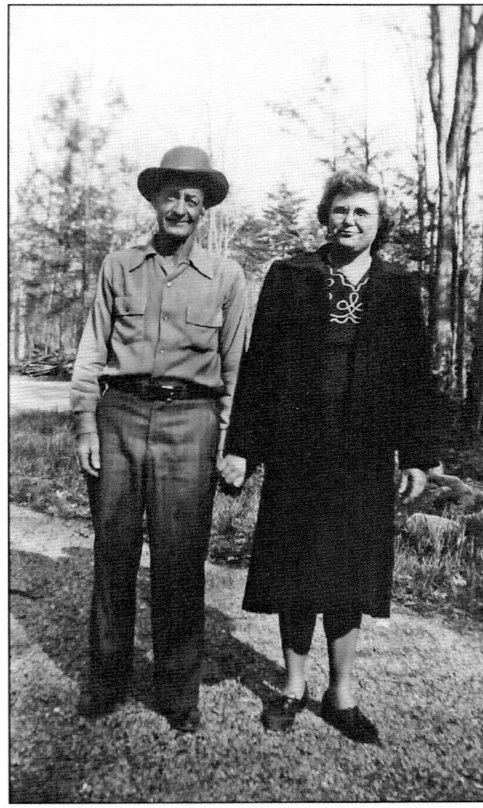

From left to right, Vina, Harley, and Otis Tipton stand behind Grace and Nola with Louise in front. They are the children of Gibson and Bertha Proctor Tipton. Gibson bought the Joe L. Walker place in Dry Valley in 1935 before moving out of Cades Cove in 1936. The Tipton family would walk over the mountain behind their house to the Tuckaleechee Primitive Baptist Church, located on Wears Valley Road. (Courtesy of Louise Tipton Loan, Bertha Proctor Tipton Collection.)

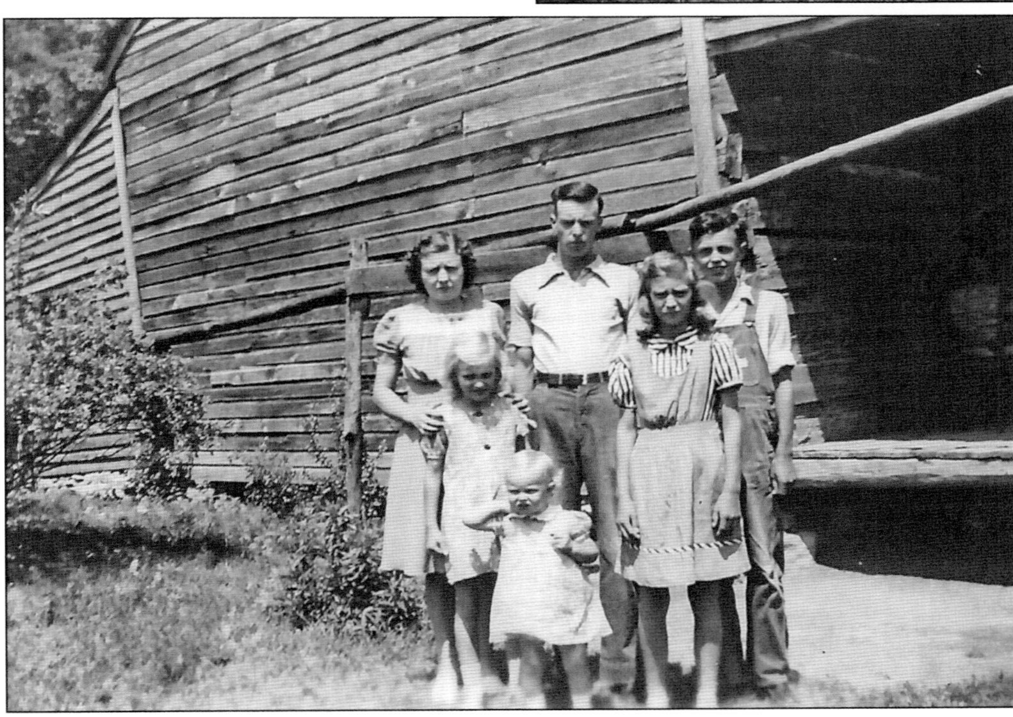

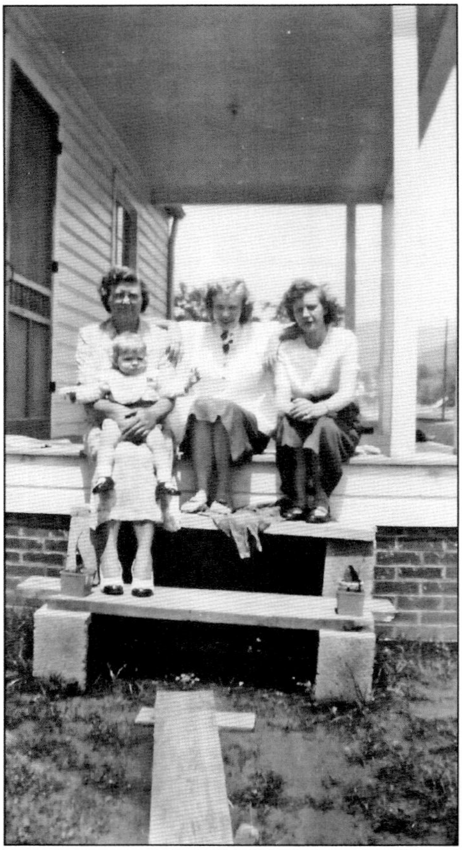

Bertha Tipton, Grace Tipton Hutsell, Nola Tipton Connor, and Donna Connor are pictured here. Prior to receiving electricity in their home, women used a washboard for scrubbing clothes, and wash water was heated over a fire. Outdoors, a clothesline was wiped clean of dirt daily, and wooden clothespins were used to hold laundry on the line. A heavy flatiron heated on the cookstove was used for ironing. (Courtesy of Louise Tipton Loan, Bertha Tipton collection.)

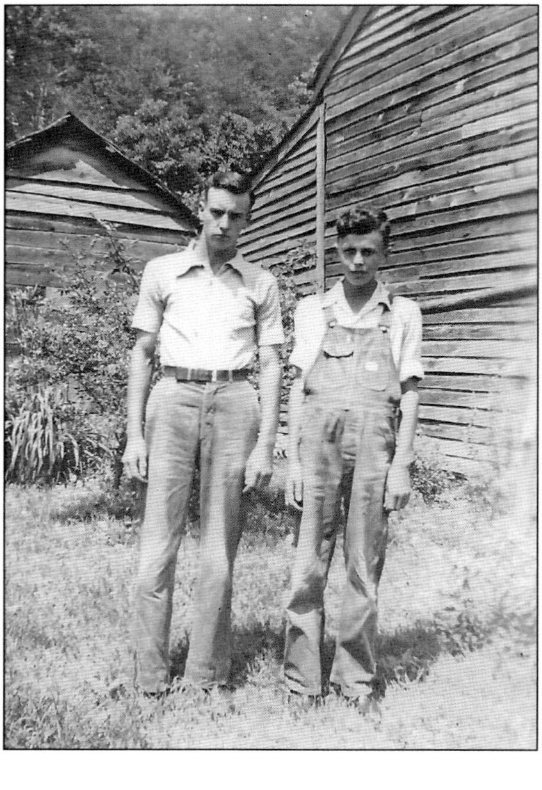

These are Harley and Otis Tipton, sons of Gibson and Bertha Proctor Tipton. As young men growing up in this era, their days likely included feeding livestock, milking the family cow, hoeing the garden, cutting firewood, attending school, and many more tasks to keep them busy. (Courtesy of Louise Tipton Loan, Bertha Tipton collection.)

Gibson Tipton and son-in-law Bill Hutsell are pictured in Dry Valley in 1959 with "Old Tolee." By this time in the 20th century, tractors had been introduced for farming. However, Gibson Tipton liked to do it the old-fashioned way. Turning the soil was likely to have taken all day. (Courtesy of Louise Tipton Loan, Bertha Tipton Collection.)

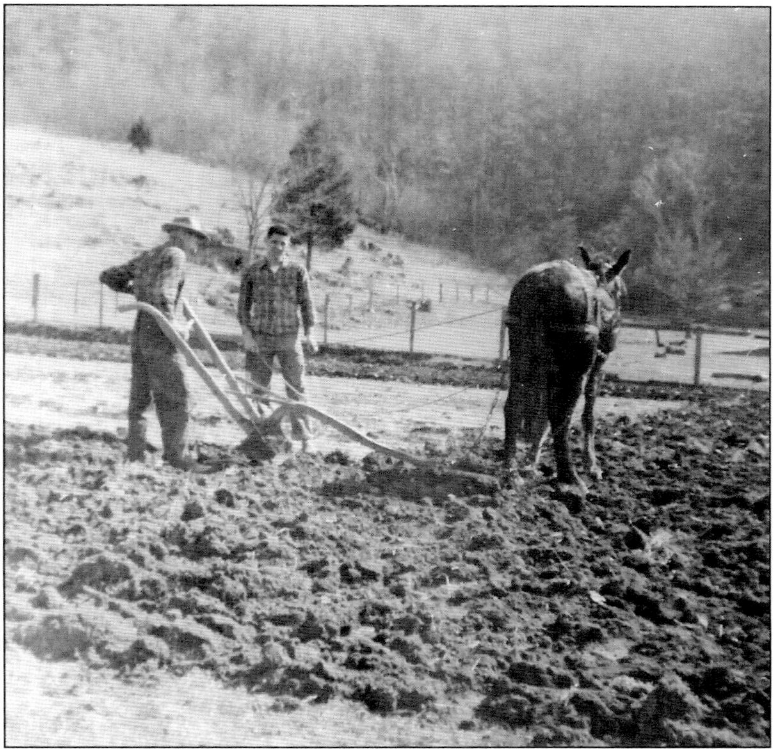

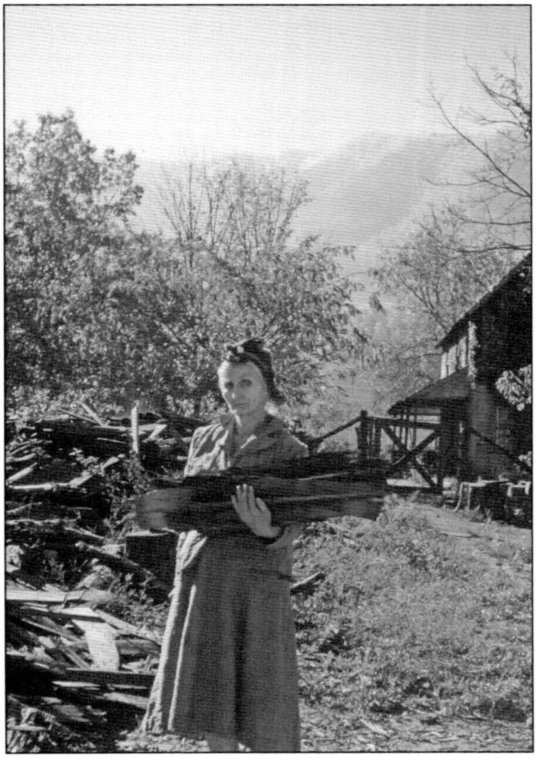

Theodocia "Doshie" Davis (1895–1981) was the wife of Sam Houston Davis and lived in Dry Valley. The Davis children—Virgie, Catherine, Geneva, Leonard, and Louise—were born and raised in the Dry Valley area of Tuckaleechee Cove. Here, Doshie is gathering wood for either her cookstove or the stove used for heating the house. (Courtesy of the William Derris Slide Collection, MS.2123. University of Tennessee Libraries, Knoxville, Special Collections.)

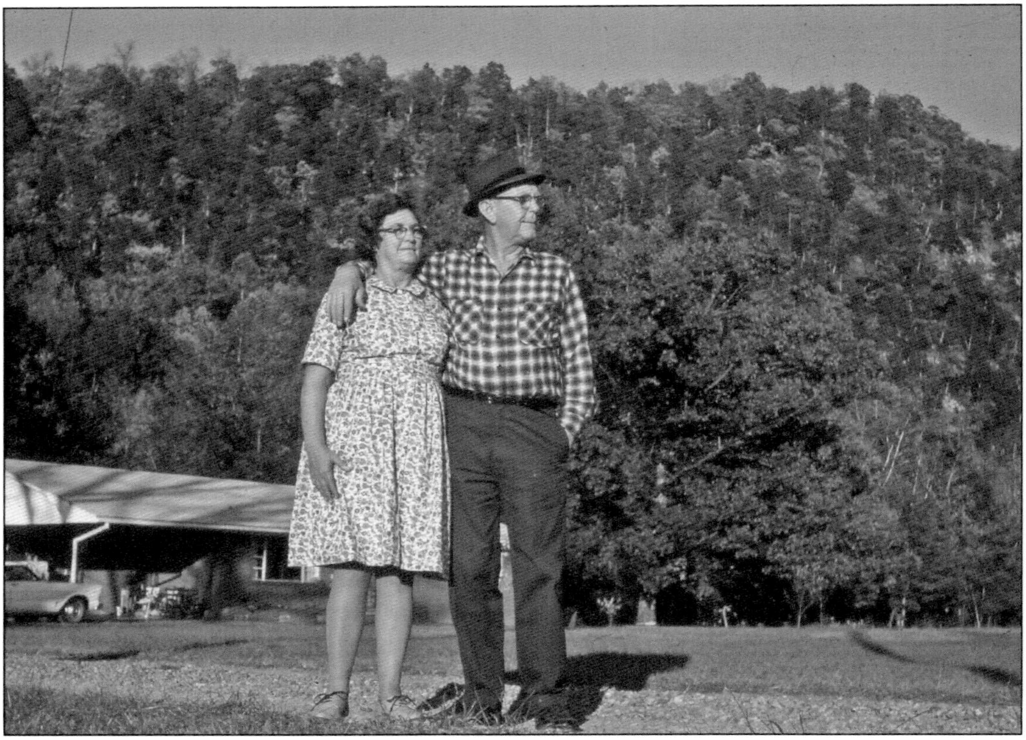

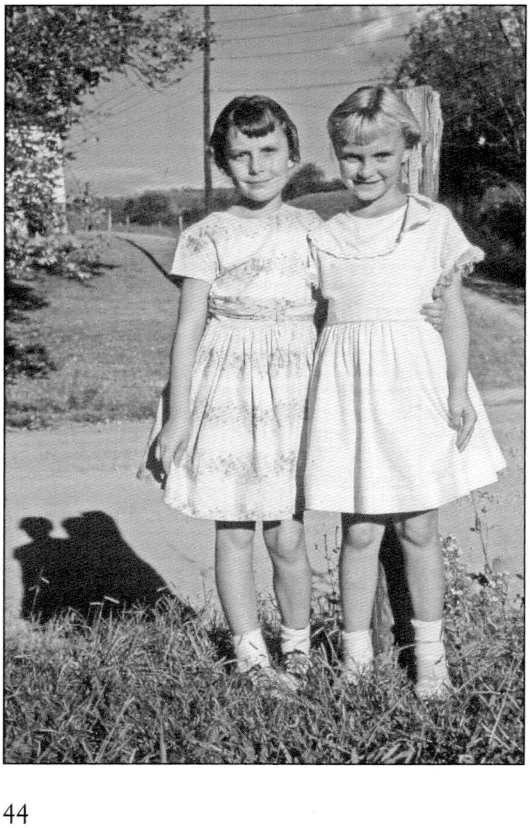

Ernest "Frosty" Myers, pictured with Mabel Hall Myers, is the son of John "Sager" and Bertha Susan Burns Myers. He and wife Mabel lived with their family in the upper end of Townsend, next to the Derris Motel. Their children are Robert, Maynard, Wilford, Dale, Colette, and Ernestine. (Courtesy of the William Derris Slide Collection, MS.2123. University of Tennessee Libraries, Knoxville, Special Collections.)

Colette (left) and Ernestine Myers are daughters of Frosty and Mabel Myers. The girls are dressed in their Sunday finest, maybe on their way to a church service or going after church for a visit to a nearby grandparent's home for Sunday afternoon dinner. (Courtesy of the William Derris Slide Collection, MS.2123. University of Tennessee Libraries, Knoxville, Special Collections.)

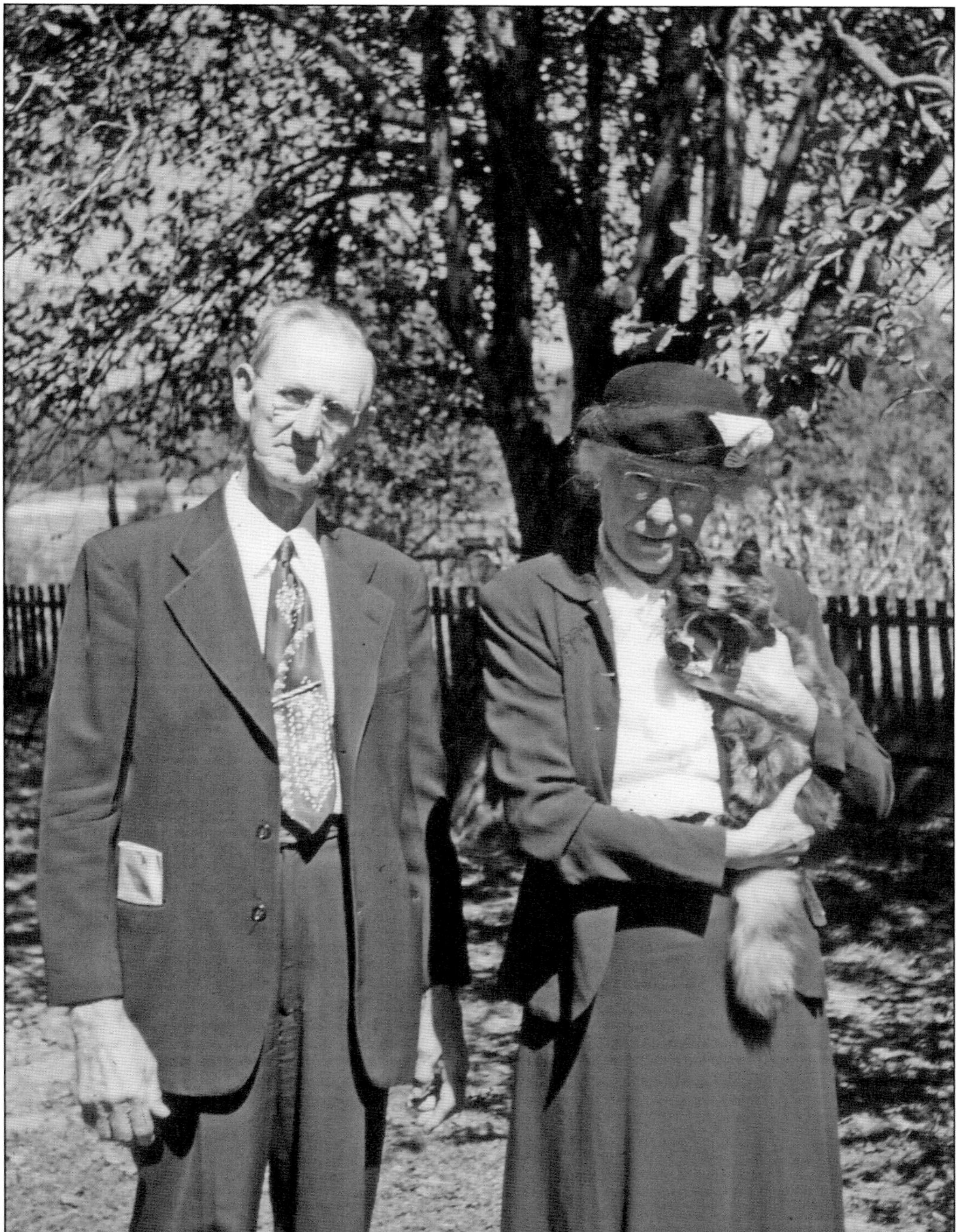

William Robinson "Bill" Dunn (September 29, 1879–February 5, 1959) and Mary Alice Rambo Dunn (August 15, 1885–June 23, 1959) are pictured. Bill was the son of William Hurst and Dorothy Snider Dunn of Tuckaleechee Cove. Mary Alice was the daughter of Daniel, a schoolteacher, and Susie Jane "Sook" Myers Rambo. Bill and Mary Alice resided in the house located beside Wears Dam. (Courtesy of the William Derris Slide Collection, MS.2123. University of Tennessee Libraries, Knoxville, Special Collections.)

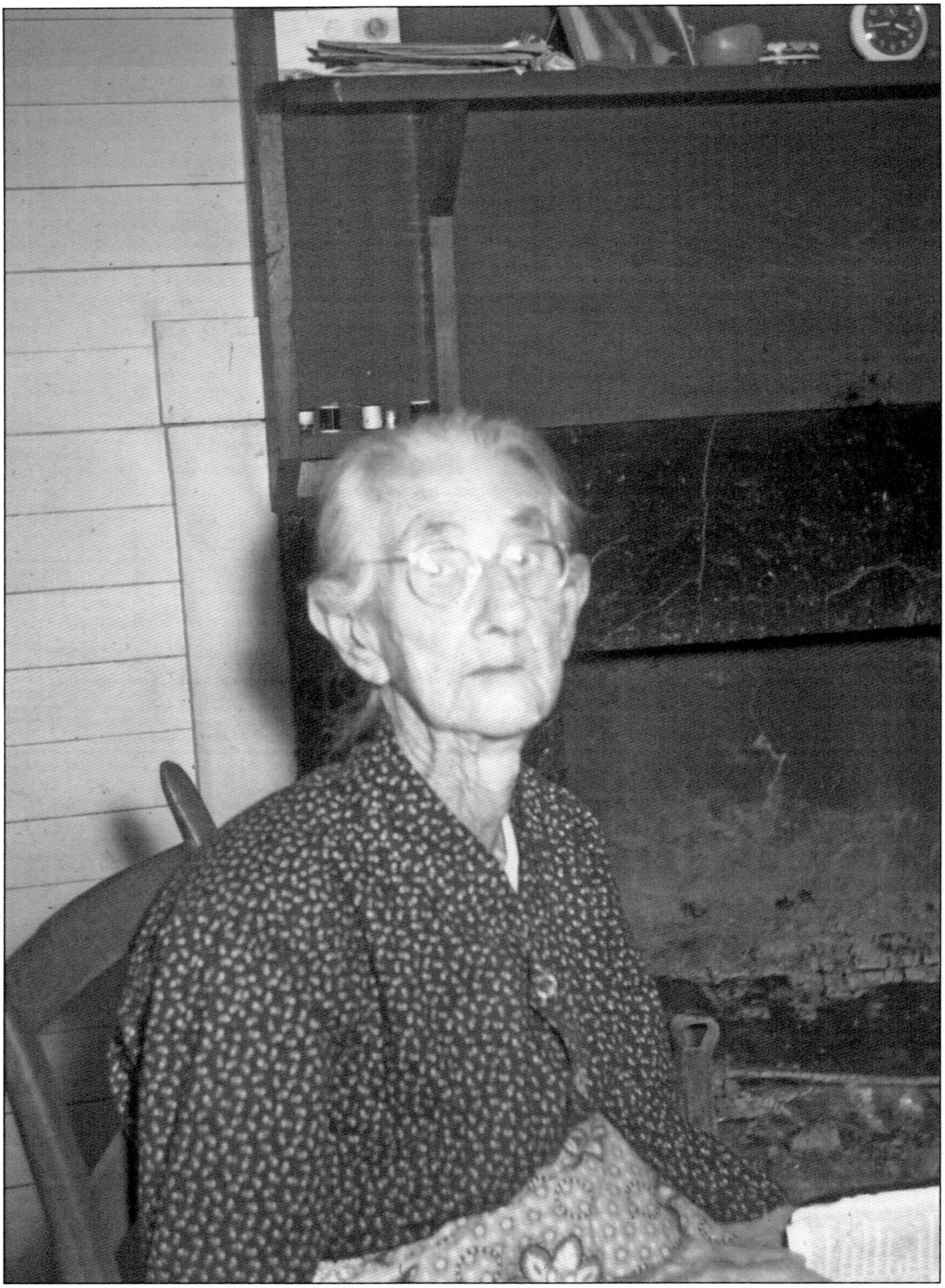

Mary Cordelia "Delia" Dunn McCampbell was born on September 12, 1870, to William Hurst and Dorothy Snider Dunn. On January 2, 1896, she married William Morris McCampbell. (Courtesy of the William Derris Slide Collection, MS.2123. University of Tennessee Libraries, Knoxville, Special Collections.)

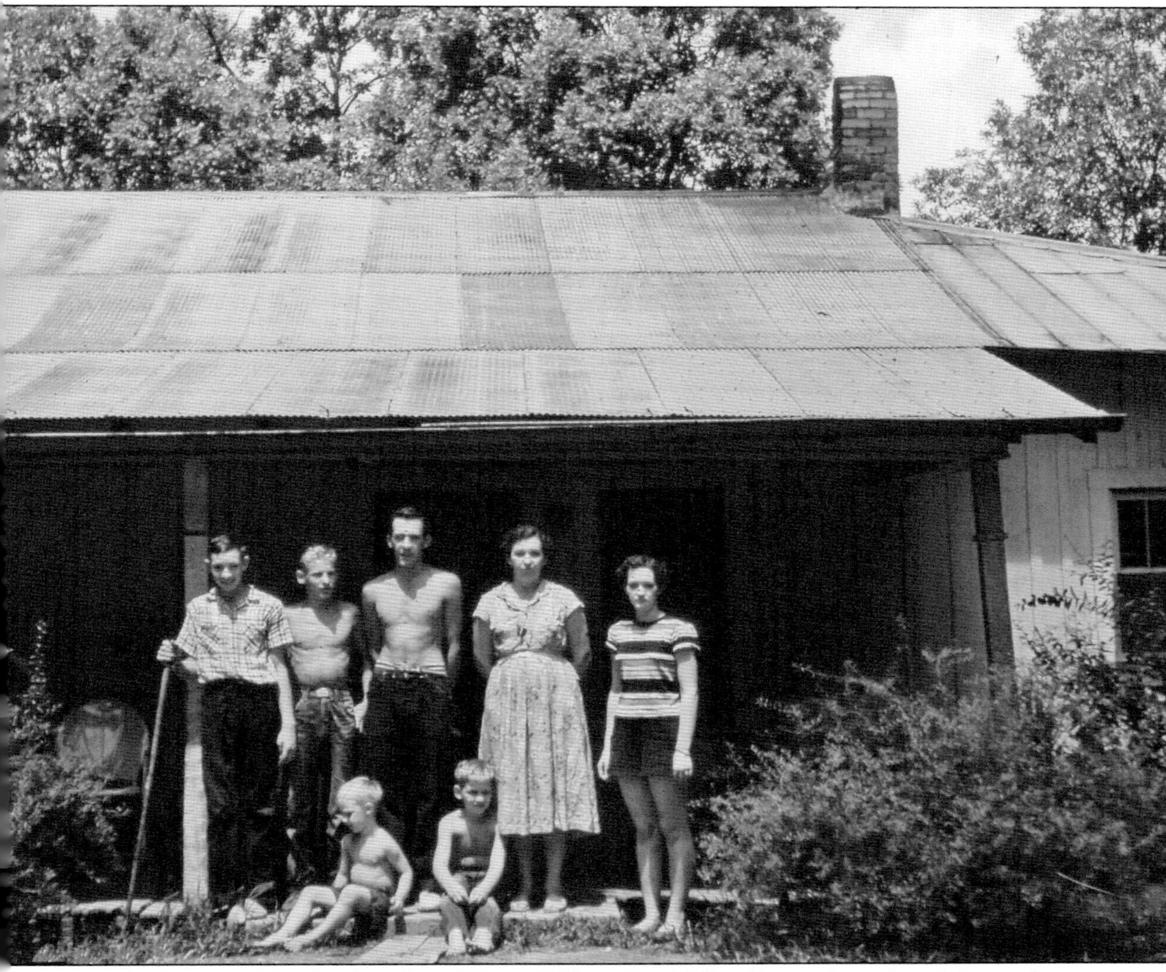

This is the home and family of Wade and Bessie Effler in Dry Valley. Pictured are, from left to right, Paul Milton, Raymond Leon, Wade, Bessie, and JoAnn Effler with Lynn Clayton and Mayford Stephen in front. Children did many daily chores, such as cutting wood, feeding livestock, and helping in garden by weeding or picking vegetables. One games that was played by the children was scotch wheel, which involved rolling a metal barrel stave around the yard with a hooked piece of wire. (Courtesy of the William Derris Slide Collection, MS.2123. University of Tennessee Libraries, Knoxville, Special Collections.)

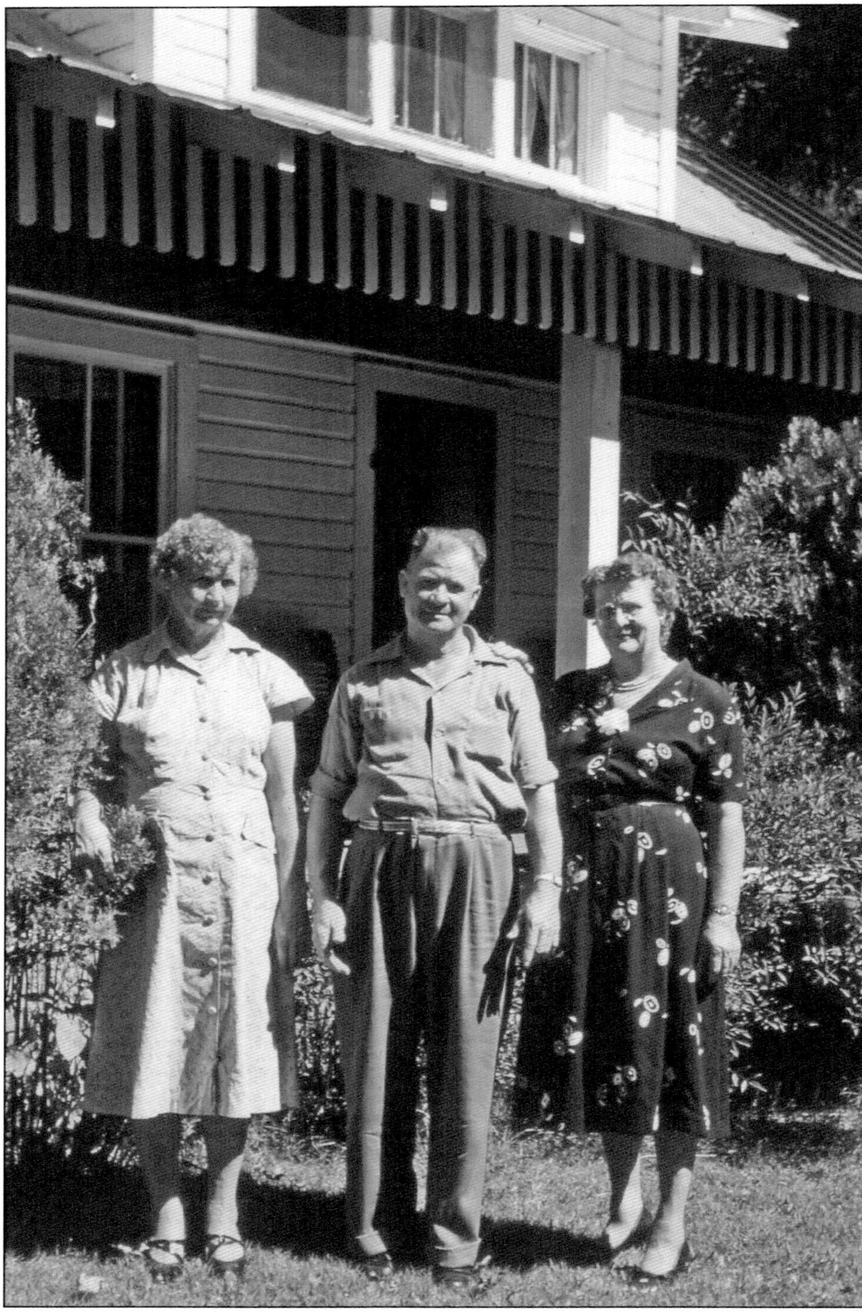

Delia (left), Peachy (center), and Josie Burns were siblings who never married and lived together in Townsend. One of Peachy's beloved pastimes was watching baseball. A favorite memory of his friends' was watching him go off to a World Series ball game. Burnses' home place was located near the dam in Townsend. The caption on this picture by the photographer stated that the photographer and the three Burnses were getting ready for a trip to listen to old harp music. (Courtesy of the William Derris Slide Collection, MS.2123. University of Tennessee Libraries, Knoxville, Special Collections.)

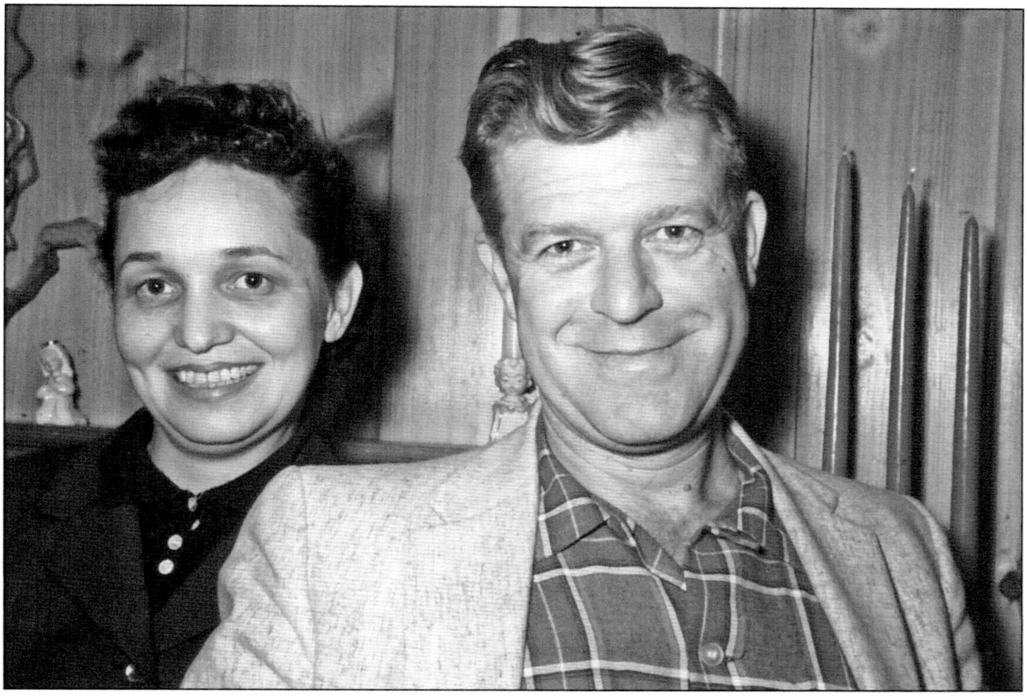

John and Norma Hitch Wilson operated Pat and Anna's Restaurant in the Kinzel Springs area of Townsend. They later built Wilson's Hillbilly Restaurant close by and ran it for many years. Located on Old Highway 73, which is now East Lamar Alexander Parkway, it was known for its tasty home cooking and hillbilly atmosphere. (Courtesy of the William Derris Slide Collection, MS.2123. University of Tennessee Libraries, Knoxville, Special Collections.)

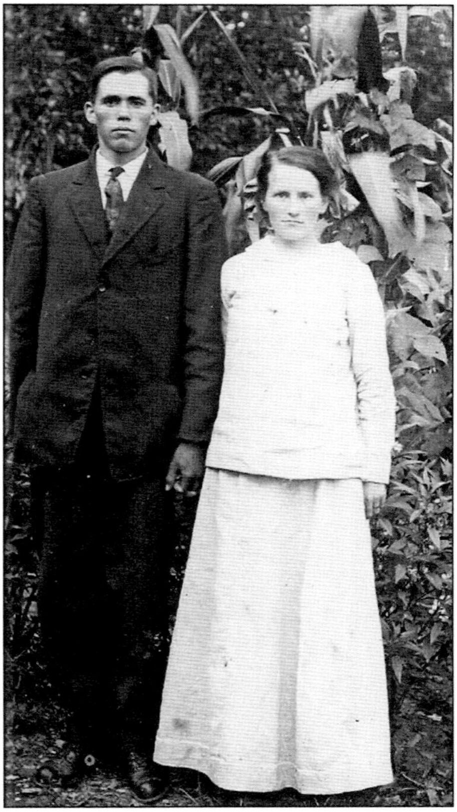

This is a photograph from the wedding of John Andrew Bryant (August 29, 1899–July 26, 1971) and Ruthie Angeline Ledbetter (July 28, 1898–May 29 1970) on September 15, 1912. John was known for making woven, white oak–splint baskets. They were used as tools, not as decoration. He made the splint by splitting white oak logs into quarters or eighths with a maul, a froe, or a mallet. The splints were separated with a knife to start then pulled apart by hand following the growth rings. This was continued until the required thinness was attained. Before the weaving process, Bryant soaked the splints in a bucket of water to make them flexible. (Courtesy of Johnnie Bryant Sparks.)

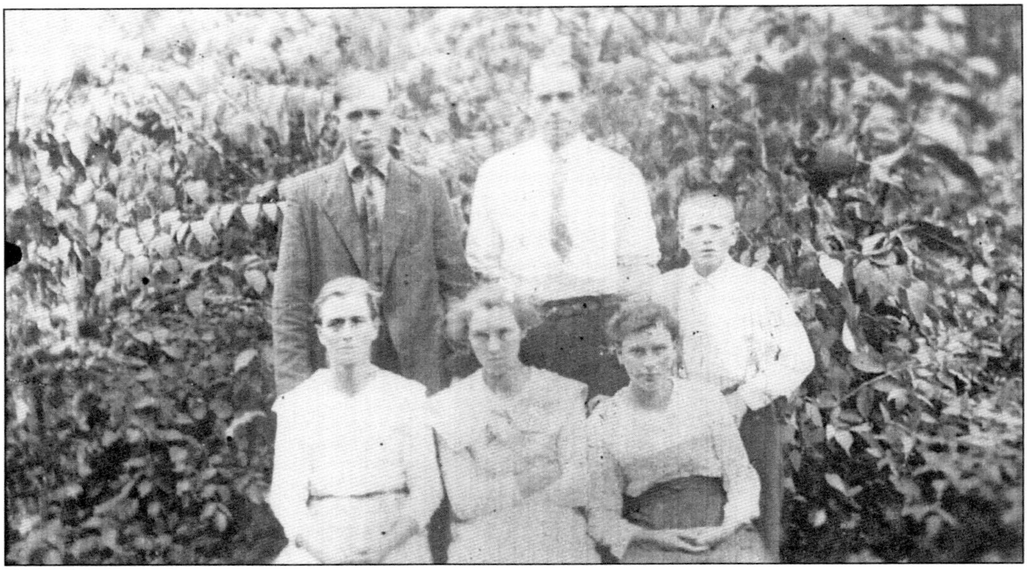

The children of Henry and Mary Shuler are, from left to right, (first row) Martha, Julie, and Mae; (second row) William, George, and Clyde. The Shuler family made their home on Rich Mountain. (Courtesy of Vance Shuler.)

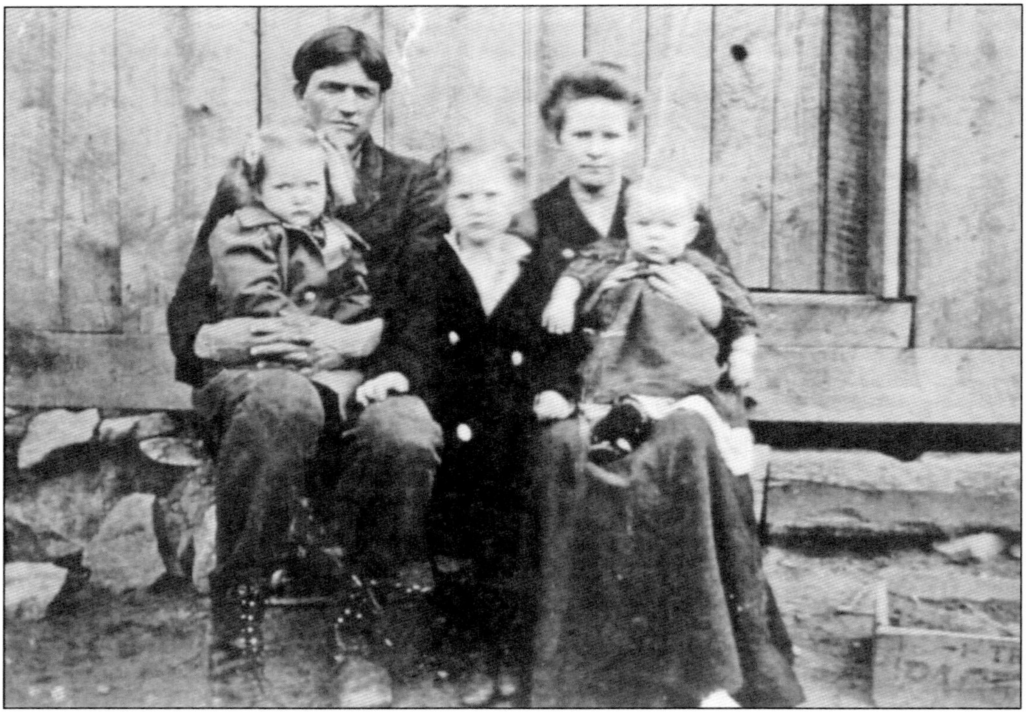

Sam and Lizzie Ledbetter Lane are pictured with their children Johnnie (left), Julia (center), and George. Lizzie was the daughter of Matthew "Bud" and Mary Ann Sands Ledbetter. (Courtesy of Paulette Ledbetter.)

Two

Religion, Education, and Sports

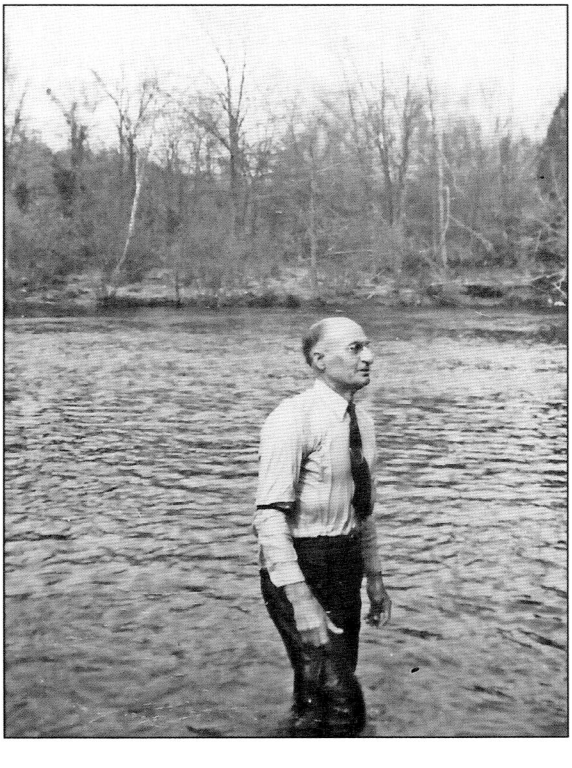

Elder Shade Tipton was a member of the Primitive Baptist Church of Cades Cove and was later ordained as minister of the Old School for Primitive Baptist. After being recommended as minister by church elders, and after being deemed sound and orthodox, Brother Shade Tipton was ordained a minister of the gospel. (Courtesy of Louise Tipton Loan, Bertha Proctor Tipton collection.)

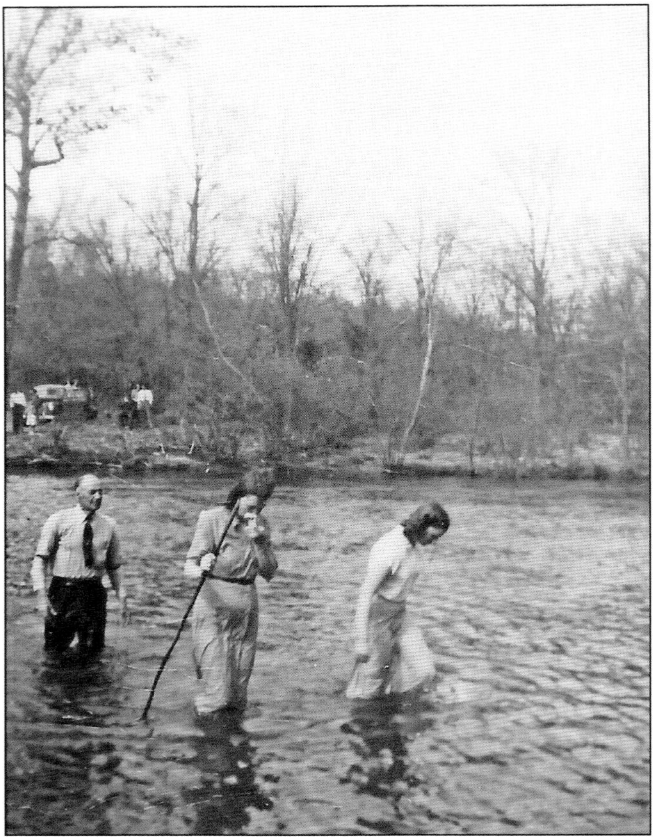

Elder Shade Tipton is baptizing Mildred and Hazel Tipton in Little River. Being baptized is the means of induction into the faith. Baptism by immersion is one of the ways to become a member of the Tuckaleechee Primitive Baptist Church. (Courtesy of Louise Tipton Loan, Bertha Proctor Tipton collection.)

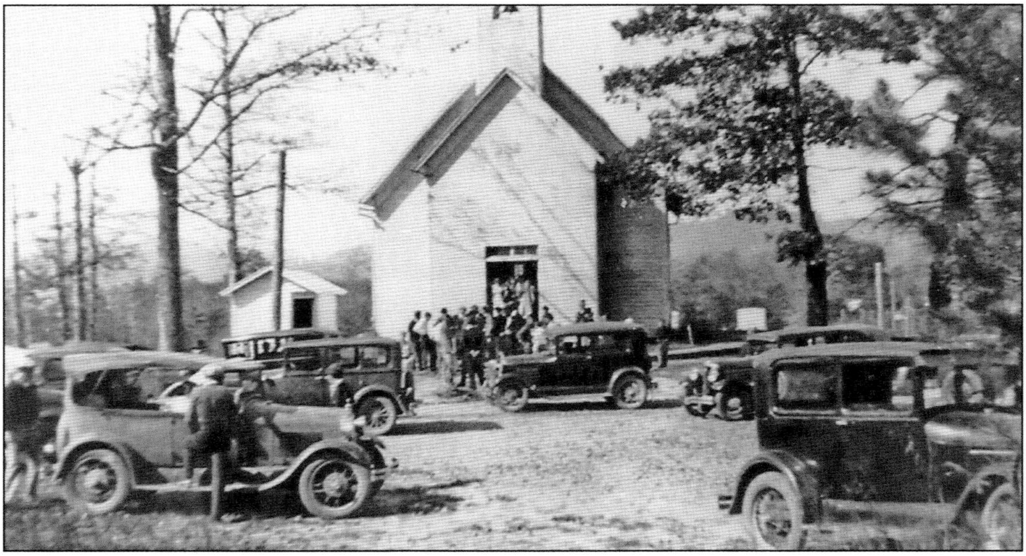

Tuckaleechee Methodist Church is pictured in 1891. Church members obtained the land from Marion Cameron and made plans to build a house of worship. The men of the church went to the mountains, cut timber, and hauled it out by horse-drawn wagons to the nearby sawmill at Dunn Springs. (Courtesy of Missy Tipton Green.)

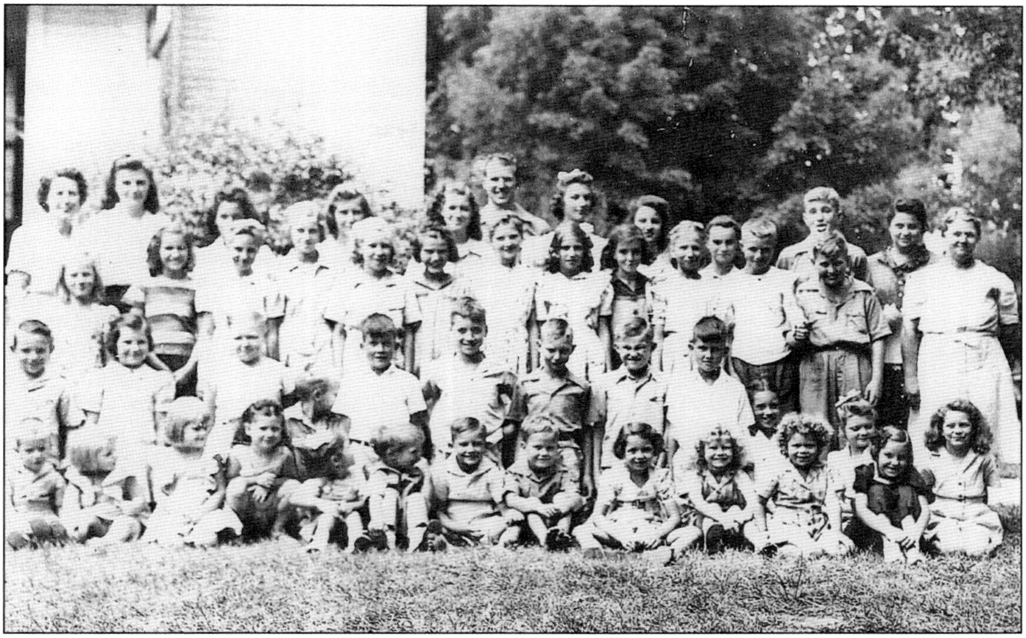

Bible school at Margaret Townsend Memorial Church is photographed here in the early 1940s. This church, named for W.B. Townsend's wife, was well attended during the years of the lumber mill, and its first service was on December 21, 1924. That same day, the building caught fire, repairs were made, and it became an active congregation. The church ceased operation at the closing of the lumber mill. (Courtesy of McNiell Family Collection.)

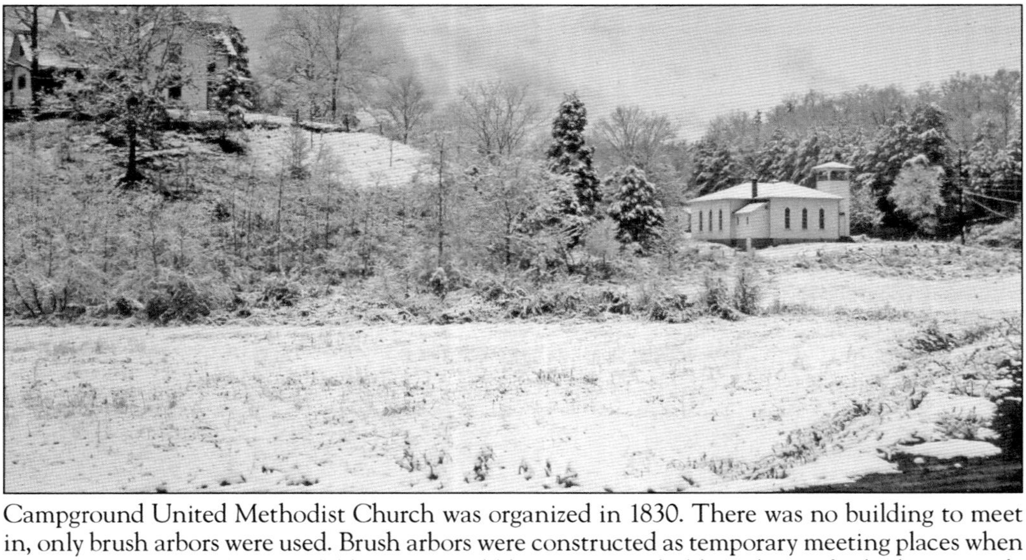

Campground United Methodist Church was organized in 1830. There was no building to meet in, only brush arbors were used. Brush arbors were constructed as temporary meeting places when there was not a building available. One would have posts to hold up the roof, which was made from different trees. Circuit-riding preachers would bring the sermons. Campground was first called Mount Pleasant or Tuckaleechee Campground. In 1869, a one-room, wood structure was built; it still stands today as part of a larger building that has undergone additions and remodeling. (Courtesy of the William Derris Slide Collection, MS.2123. University of Tennessee Libraries, Knoxville, Special Collections.)

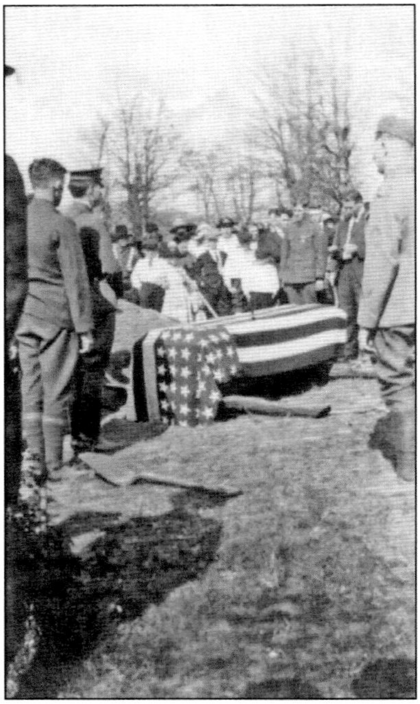

This is the funeral of William "Pruitt" Myers, son of Joseph and Laura Feezell Myers. Pruitt, who was born on December 25, 1889, was killed in action on October 7, 1918, in Geneve, France during World War I; he was 28 years old. He was survived by his wife, Rettie Tipton Myers. It took two years for his body to arrive back in Townsend; it came by train along with the body of his close friend Henry T. Webb to the Little River Lumber Mill depot and was then taken by wagon hearse to Bethel Baptist Church for a double funeral. (Courtesy of McNiell Family Collection.)

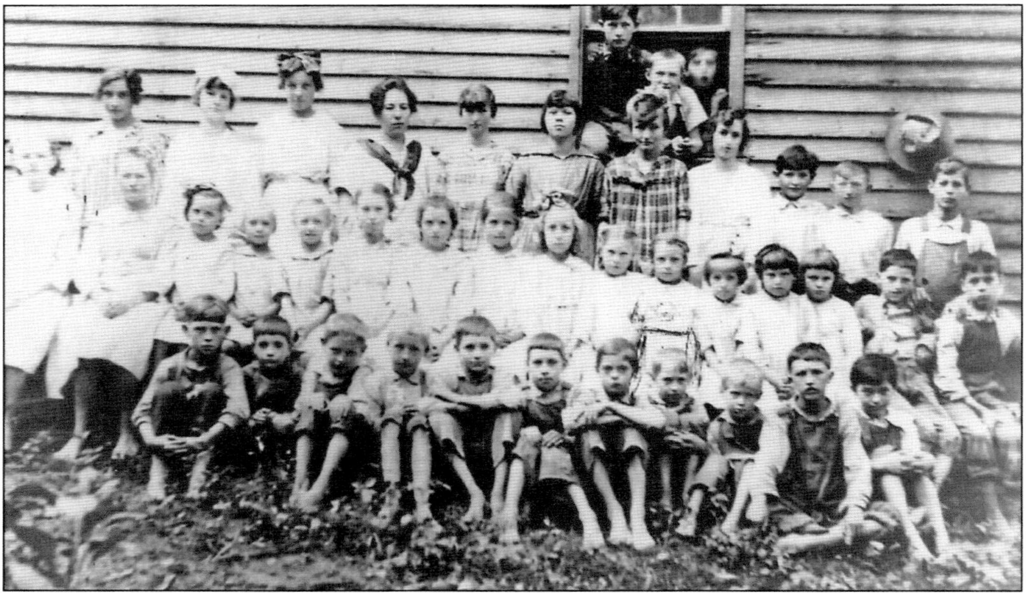

This 1919 image was taken at Piney Hill School, also known as Caylors School. In the window are Clark Scott (left), Clay Webb (center), and Earnest Caylor. Outside the building are, from left to right, (first row) Harvey Woodby, Amos Caylor, Earl Dunn, Lee Caylor, Witt Caylor, Edgar Caylor, Wade Caylor, Lloyd Caylor, Luther Caylor, Harry Ferguson, and Y.J. Morris; (second row) Ida Caylor, Pearl Caylor, Sylvia Street, Ruby Cotter, Louella Cotter, Zella Caylor, Eva Caylor, Maggie Caylor, Leona Keeler, Hazel Abbott, Cecil Keeler, Irene Keeler, Jennie Woodby, Ola Williams, Herald Ferguson, and Harvie Caylor; (third row) Bessie Caylor, Vina Woodby, Grace Feezell, teacher Willie Davis, Lela Caylor, Golden Williams, Angie Caylor, Edna Ferguson, Leonard Scott, Frank Davis, and Perry Caylor. (Courtesy of Paulette Ledbetter.)

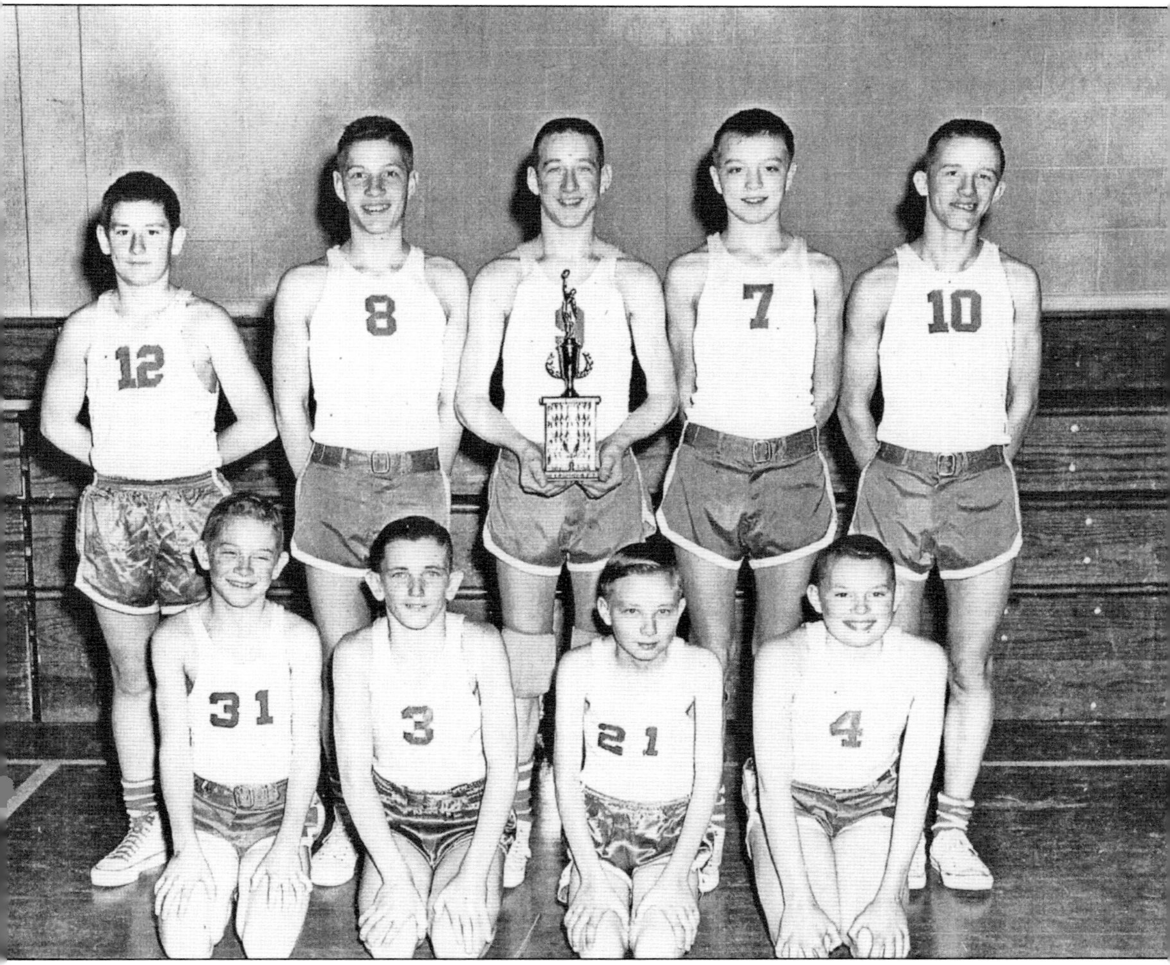

The players of the Townsend Elementary basketball team were, from left to right, (first row) Chuck Hannah, David Murrell, Leon Brickey, and Tommy Lail; (second row) Tim Stinnett, Steve Vananda, Joe Myers, David Compton, and John Huskey. (Courtesy of Townsend Alumni Photograph Collection.)

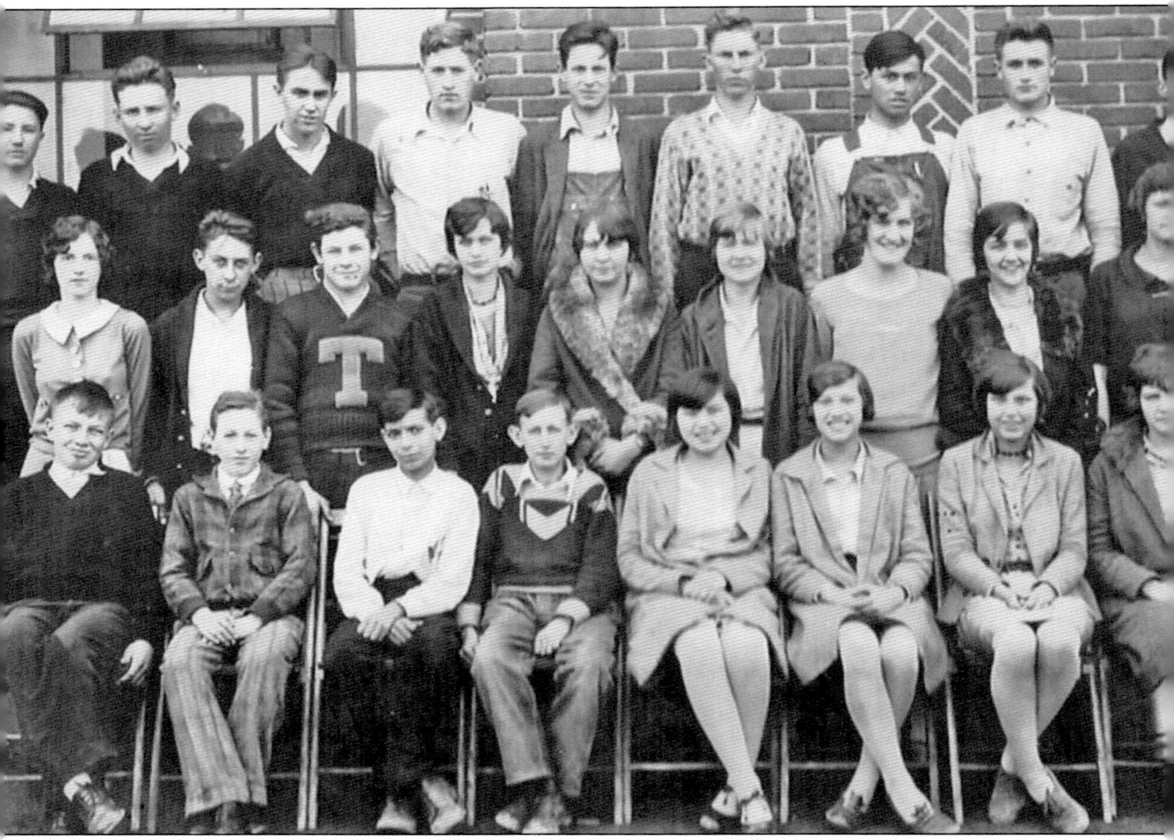

In this photograph of the Townsend High School class of 1929 are, from left to right, (first row) Raymond Walker, Calvin Dunn, Ray Lawson, Herbert Keeler, Ada Lawson Less, Katherine Dunn Harrison, unidentified, and Mary Helen Conner Abbott; (second row) sponsor Peg Rowan Anderson, Sam Headrick, Hubert Burns, Florence Lawson Webb, Ada Florence Lawson McPherson, Helen Wear Talley, Rose Davis Gregory, Maude Walker Tipton, and Bessie Maples; (third row) Willie Myers (from Cades Cove), Charlie Burns, Johnny Phillips, Hurst Davis, Perry Shields, Paul Lawson, Isaiah Skidmore, and Clay Davis. (Courtesy of Marilyn Myers Byrd Photograph Collection.)

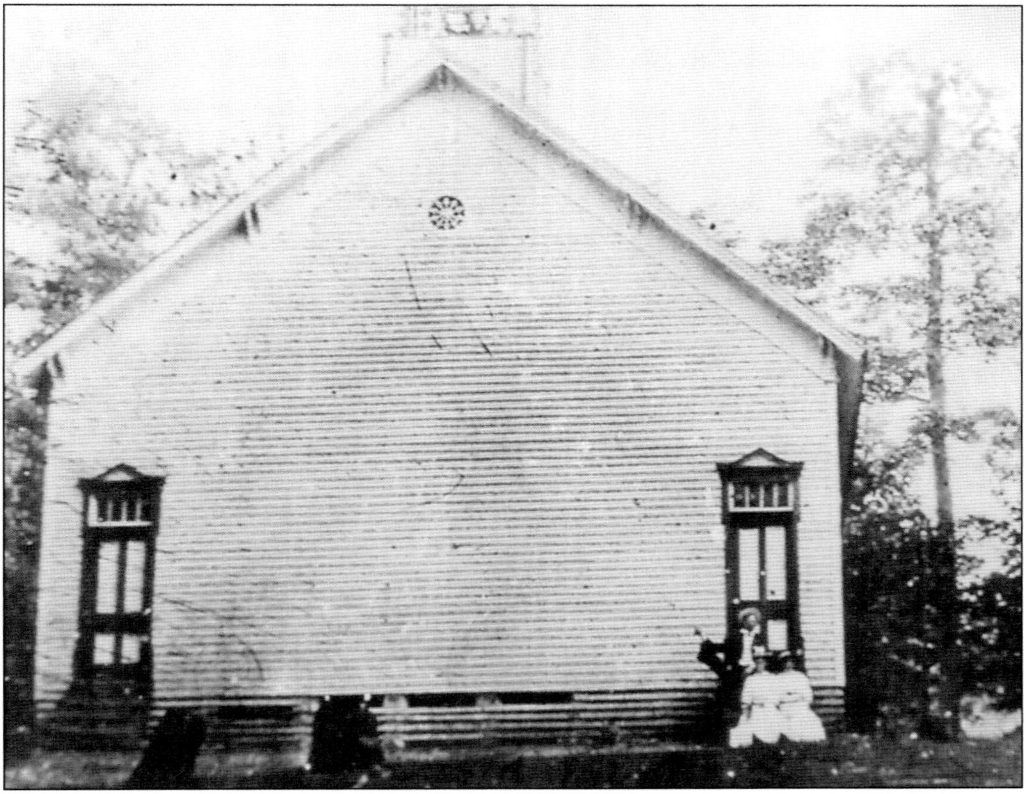

In 1879, a new church was built on Old Highway 73 up the hill from an older church that stood closer to the road. In 1881, it was dedicated, and the name was changed to Bethel Baptist. During the years, many things have been added to Bethel, including a belfry, Sunday school rooms, and electricity. The old frame building was torn down in 1954 so a new church could be constructed. The brick structure still stands today. (Courtesy of Missy Tipton Green.)

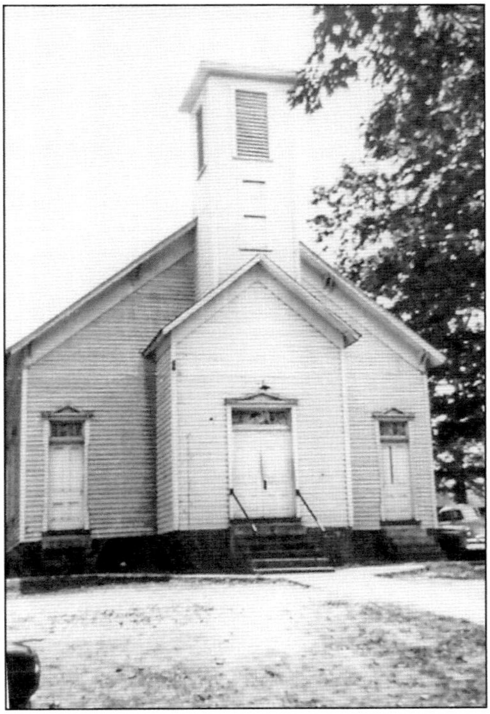

The Baptist church was organized in 1803 as the United Baptist Church of Christ of Tuckaleechee Cove. No written records could be found for the period 1810 to 1835. Around 1836, the church started up again, but membership dropped as some of the members left to start another church, the Primitive Baptist Church. In 1840, the mission side of the church began using a school as a place of worship. Around 1860, construction of a new church started. The Civil War was a time when many people did not attend church. (Courtesy of McNiell Family Collection.)

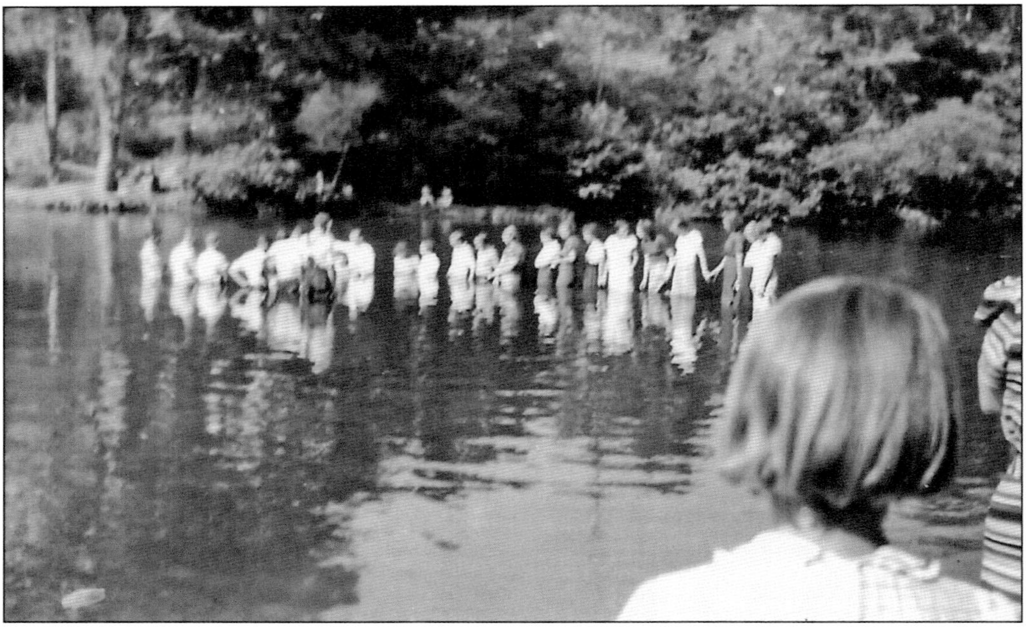

Bethel Baptist Church held its baptisms at the McKelder hole (or McKeldry hole) on the banks of Little River. At times, being immerged in the cold waters of Little River could be rather frigid. The attending congregation would sing "Shall We Gather at the River." Baptismal services were a very inspiring time. (Courtesy of McNiell Family Collection.)

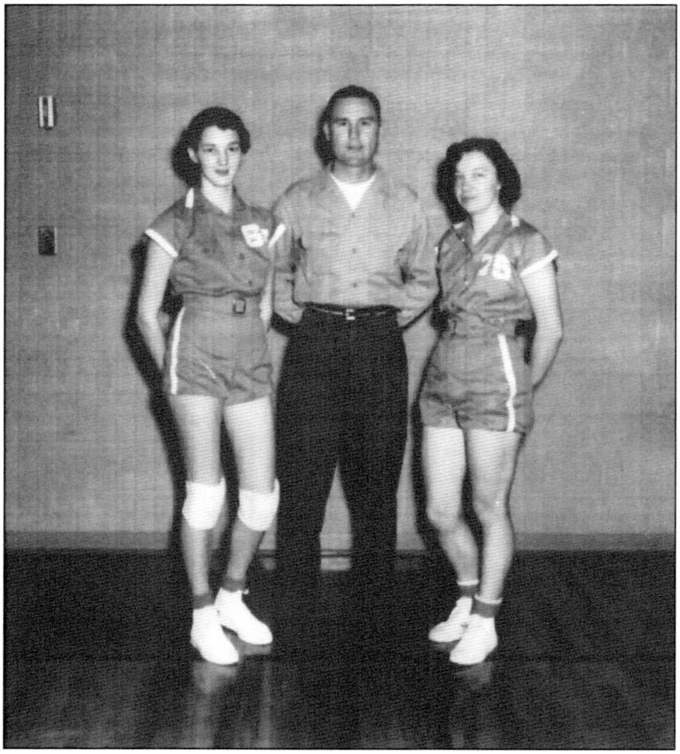

Thomas "Honk" Boyd came to coach the Townsend High School girls' basketball team in the early 1950s. Everyone said the "Honk" was an excellent coach, and the team enjoyed playing for him. Pictured from left to right are Betty Lawson, Thomas "Honk" Boyd, and Betty Walker. An incentive for basketball players, football players, and baseball players were to receive a letter "T" to wear on either their school jacket or sweater. This was an accomplishment for players, and they were proud to wear the jacket. After they received the letter, for the following years they played they received a stripe or chevron to be added to the jacket or sweater. (Courtesy Townsend Alumni Photograph Collection.)

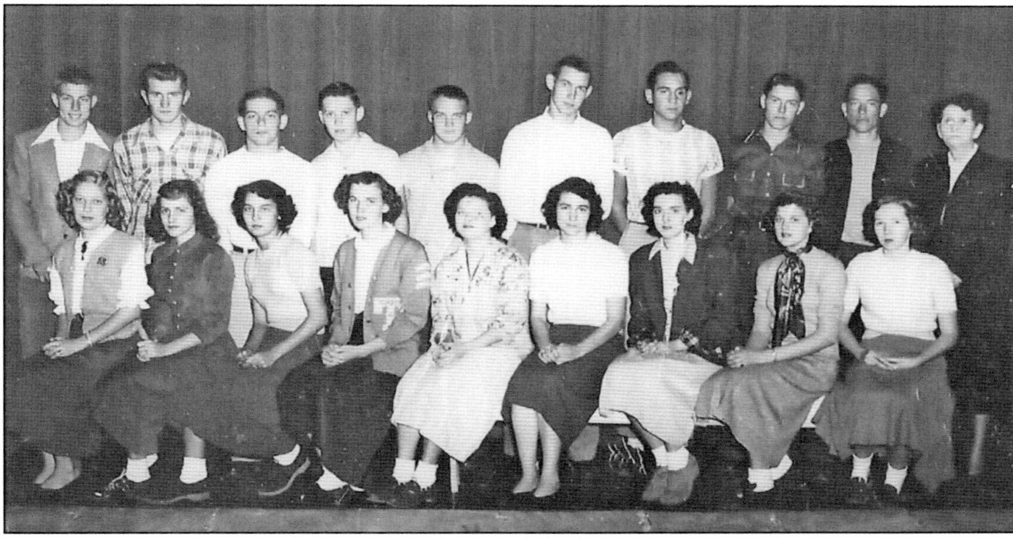

This is the Townsend High School graduating class of 1950. From left to right are (first row) Betty Stratton, Coleen Emert, Alda Rathbone Bricky, Bennie Webb, Nellie Cooper, Lenore Sparks, Catherine Tipton, Billie June Lawson, and Lois Myers; (second row) Dale Carnes, Jack Webb, Bill Effler, Van Myers, Rex Woodby, Carl Skeet Myers, J.C. Everett, Charles Myers, Edward Walker, and teacher Eleanora Adams. Reaching the important graduation day was a milestone for the Townsend students. A ceremony was held for those graduating from eighth grade as well as twelfth grade. These events were attended by all family and friends. (Courtesy of Marilyn Myers Byrd Photograph Collection.)

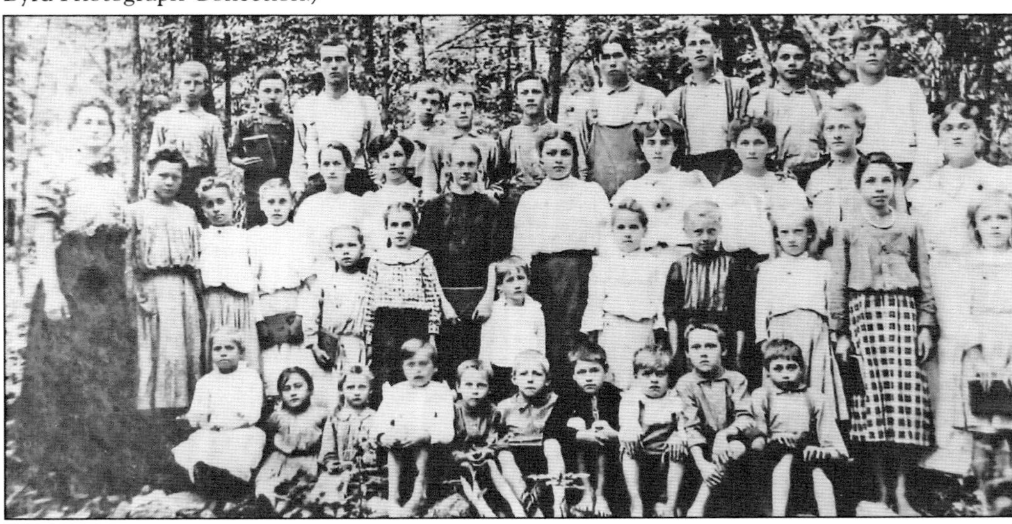

This group from Red Bank School is pictured in 1910. In the photograph are, from left to right, (first row) Eva Tipton, unidentified, Essee Gregory, Sam Handley, Willie McCampbell, Wade Tipton, Houston Effler, unidentified, and Claude Bryant; (second row) three unidentified individuals, Lillie Effler, Annee McCampbell, Ethel Lawson, Lora Stephenson, and Rose Lawson; (third row) two unidentified individuals, Bertha Dunn, unidentified, Alice McCampbell, Lora McCampbell, Queenie Dunn, and Jane McCampbell; (fourth row) Earl H. McCampbell, Mel Stephenson, four unidentified boys, John Bryant, Charlie Bryant, and Larry Handley. (Courtesy of Joann Effler Shuler Collection.)

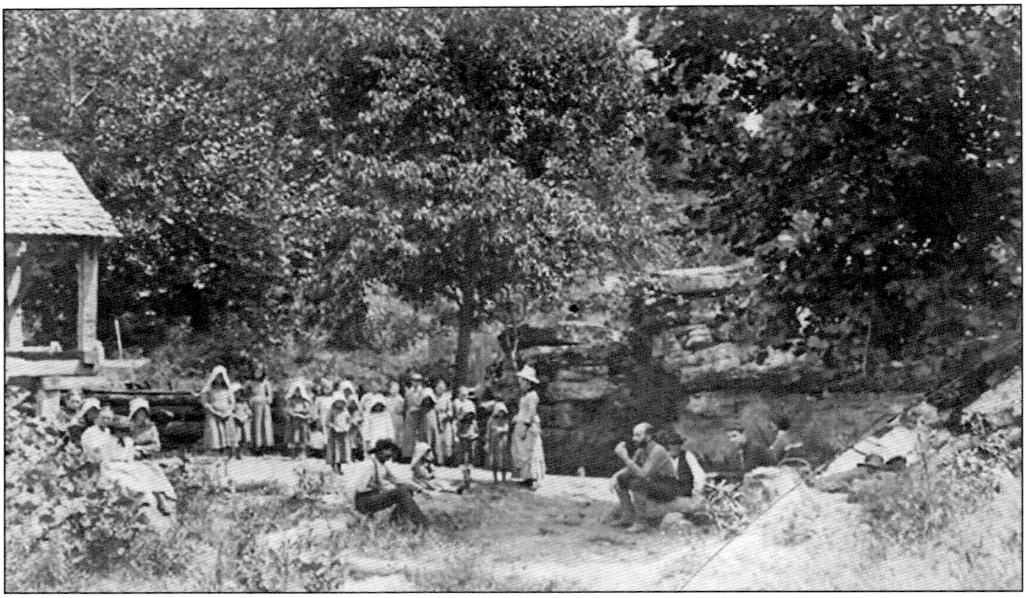

This school group is taking a lunch break at McCampbell Mill in the Dry Valley area. This was taken around 1886, so these could have been students of the College Hill School, which was in Dry Valley. (Photograph by W.C. Cochran, courtesy of the University of Tennessee Libraries, Knoxville, Special Collections.)

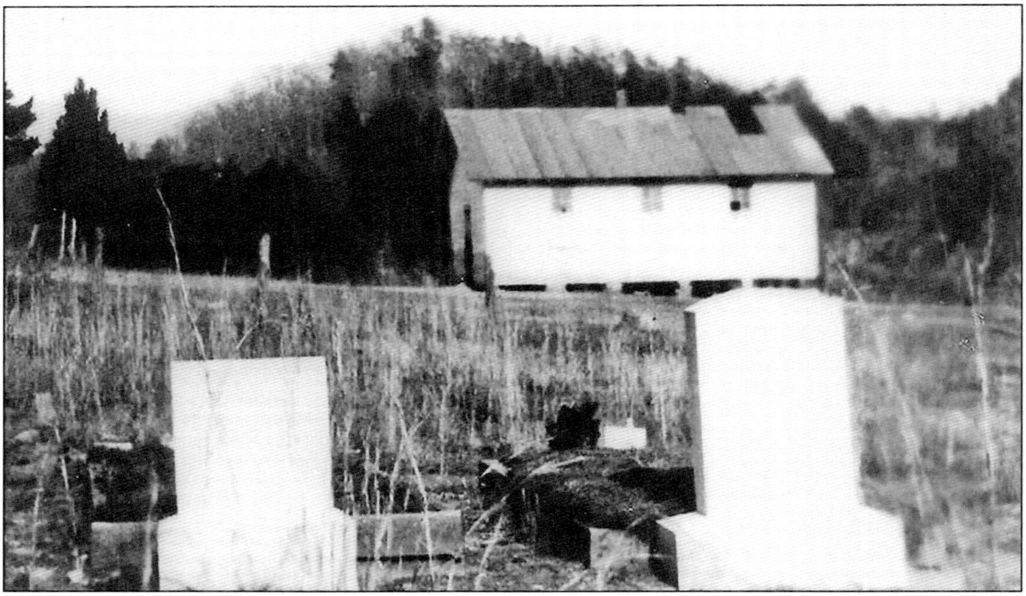

This is Caylors School, also known as Piney Hill School. Church and school activities were usually linked together. To attend Caylors School, a child had to be at least six years of age. Students had to walk to and from Caylors School, which could sometimes take an hour in the mornings and an hour in the evenings. Lunch was eaten under a large tree in the schoolyard. Baskets of food prepared by the children's mothers were spread out for the students to eat together. The restroom for the school was the largest bush the children could get behind. (Courtesy of Daniel Davis Collection.)

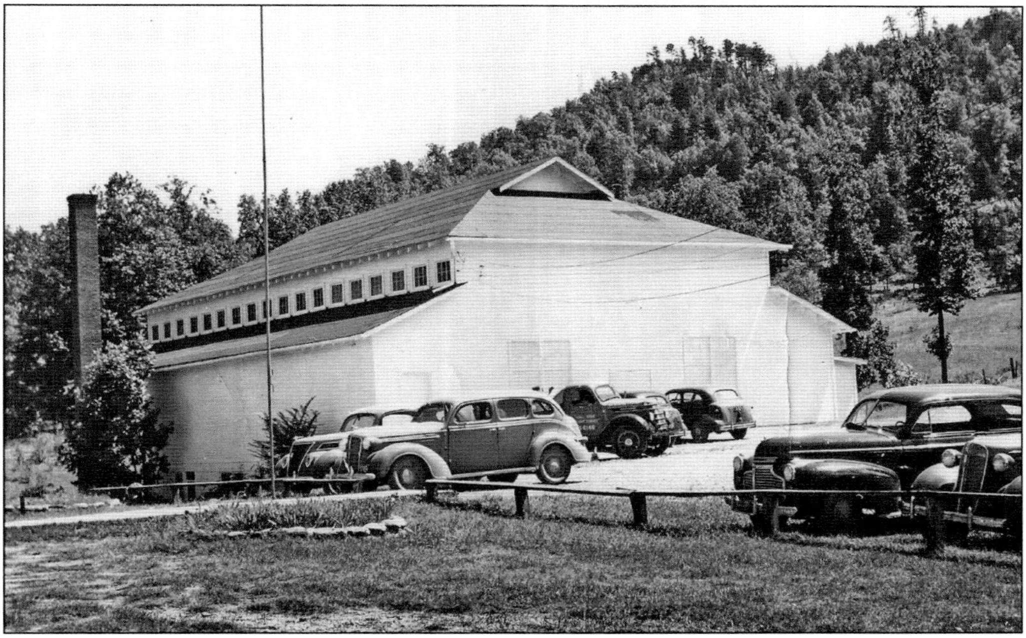

This is the Townsend gymnasium with the lunchroom to the rear. In the late 1940s and early 1950s, the federal government began a free-lunch program for families that were in need. Before the school had a lunchroom, all children had to bring their lunch to school. A new gymnasium was built in 1953. (Courtesy of Townsend Alumni Photograph Collection.)

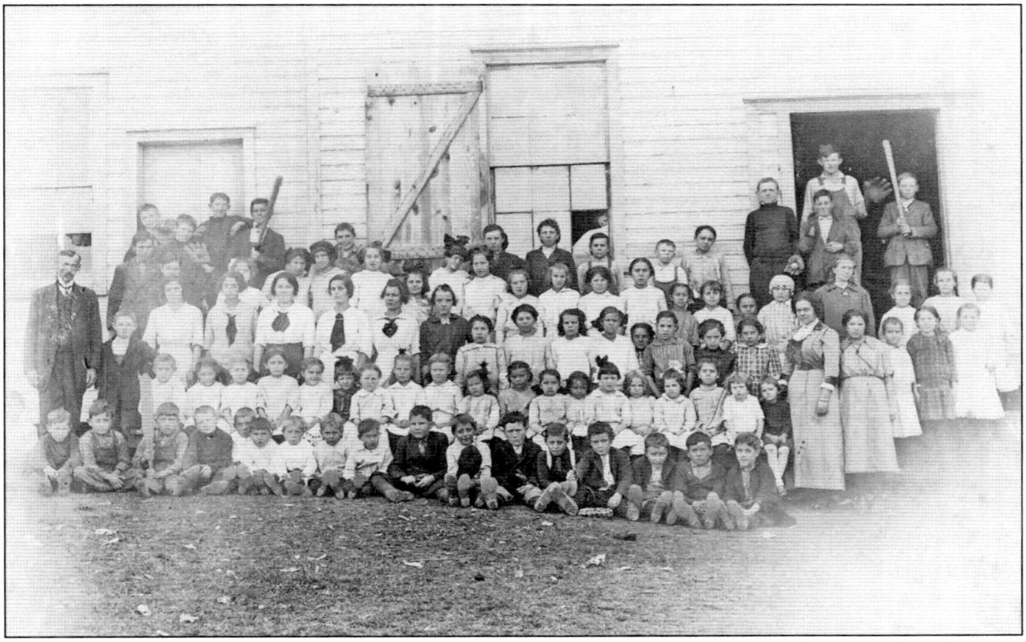

These are schoolchildren at the Townsend Elementary School in 1909. This institution was also known as the George Emert School after its principal and teacher. Sports were an important part of school life; in this picture, the students have handmade bats and what look like commercial ball gloves. (Courtesy of Marilyn Myers Byrd Photograph Collection.)

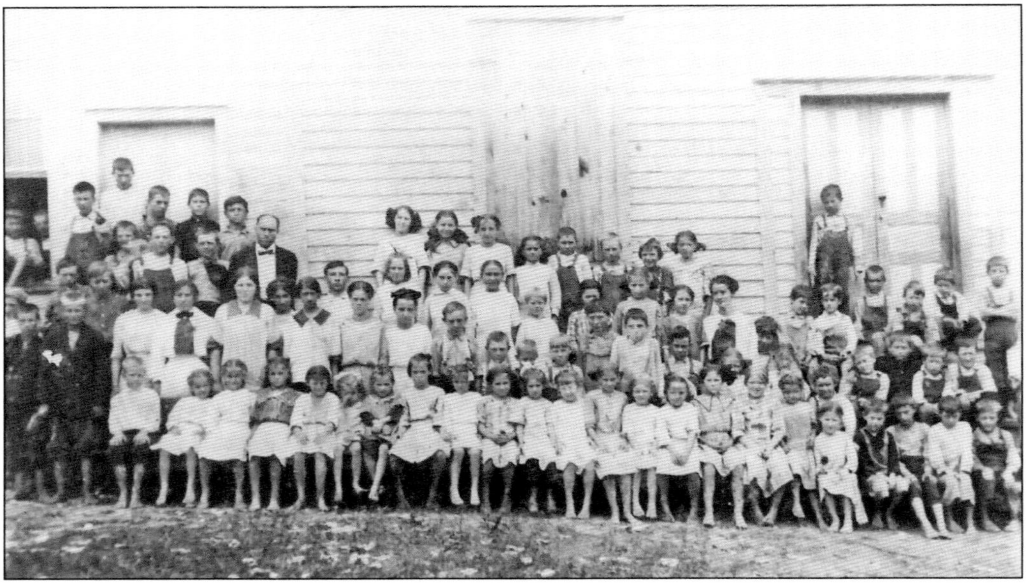

The Townsend Elementary School was a white board building that sat on the hill across from the high school. This structure accommodated grades one through five. Coal stoves were used to heat it in the colder months, and the custodian would get up early and come in to get the building warm. By 1948, there was a need for a larger elementary school due to the closures of nearby Caylors and Cables schools. The new brick building held grades one through eight and had a lunchroom on the ground level. When one had to go to the bathroom, one had to travel outside to get there. (Courtesy of Marilyn Myers Byrd Photograph Collection.)

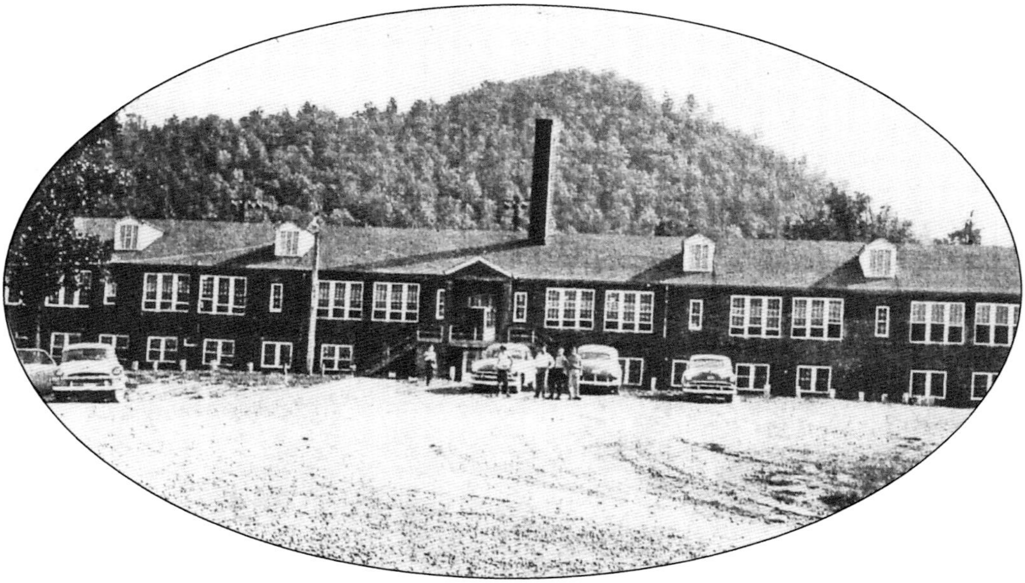

For many years, the schools in rural areas of Blount County did not have libraries. Teachers traveled to nearby Maryville to check out books at the central school office. Around 1959, elementary schools received libraries, and the Townsend Parent-Teacher Association helped purchase books. Students bought their own textbooks; the use of free textbooks did not occur until later on. (Courtesy of Missy Tipton Green.)

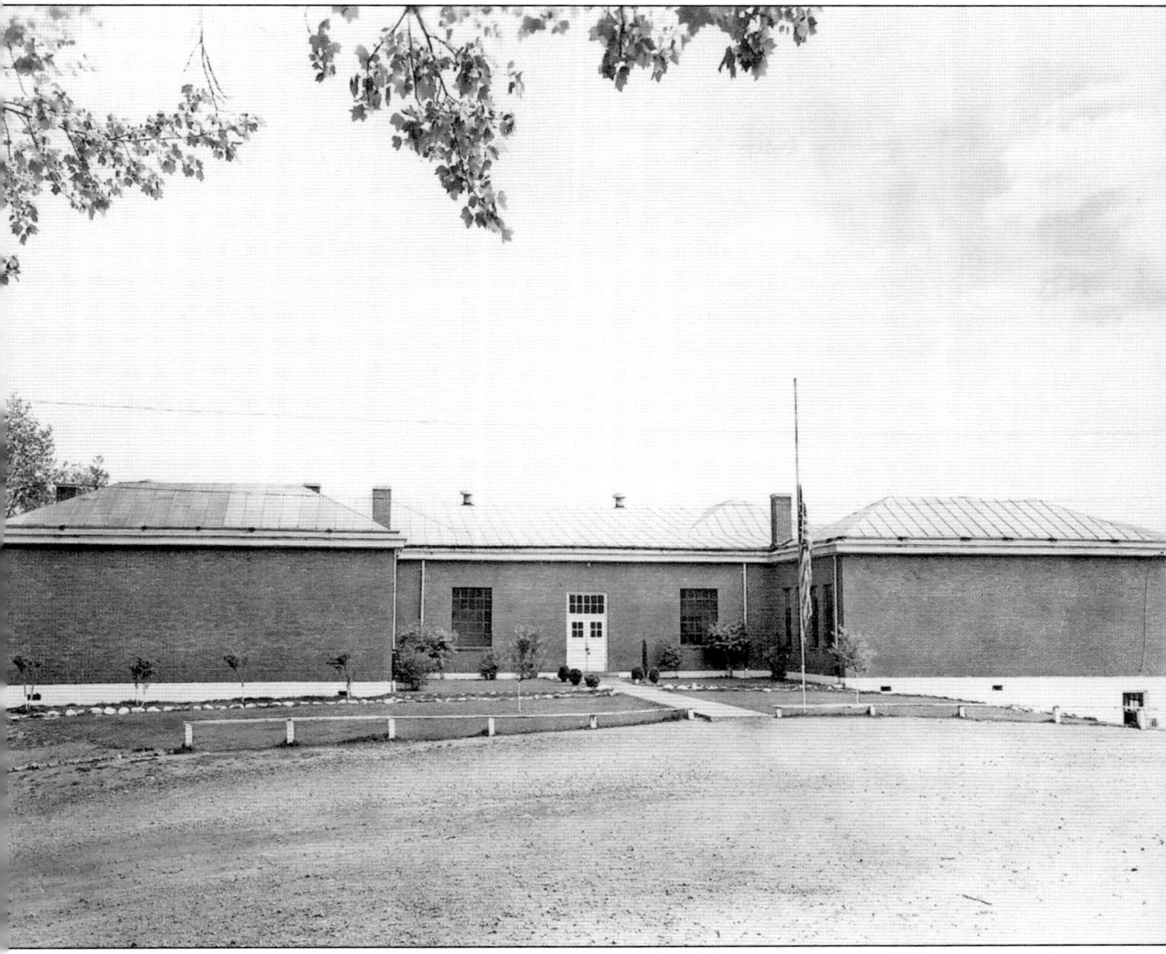

Townsend High School began in 1923 as a one-year high school. It was named for W.B. Townsend, who furnished materials and labor to build the seven classrooms and auditorium building. Classes were added each year as students achieved the next grade. Little River Lumber Company provided a supplement to fund a nine-month school term and paid for the salary of some of the teachers. The high school had indoor restrooms. Teachers could count on the support of the parents and were well respected in the community. (Courtesy of Missy Tipton Green.)

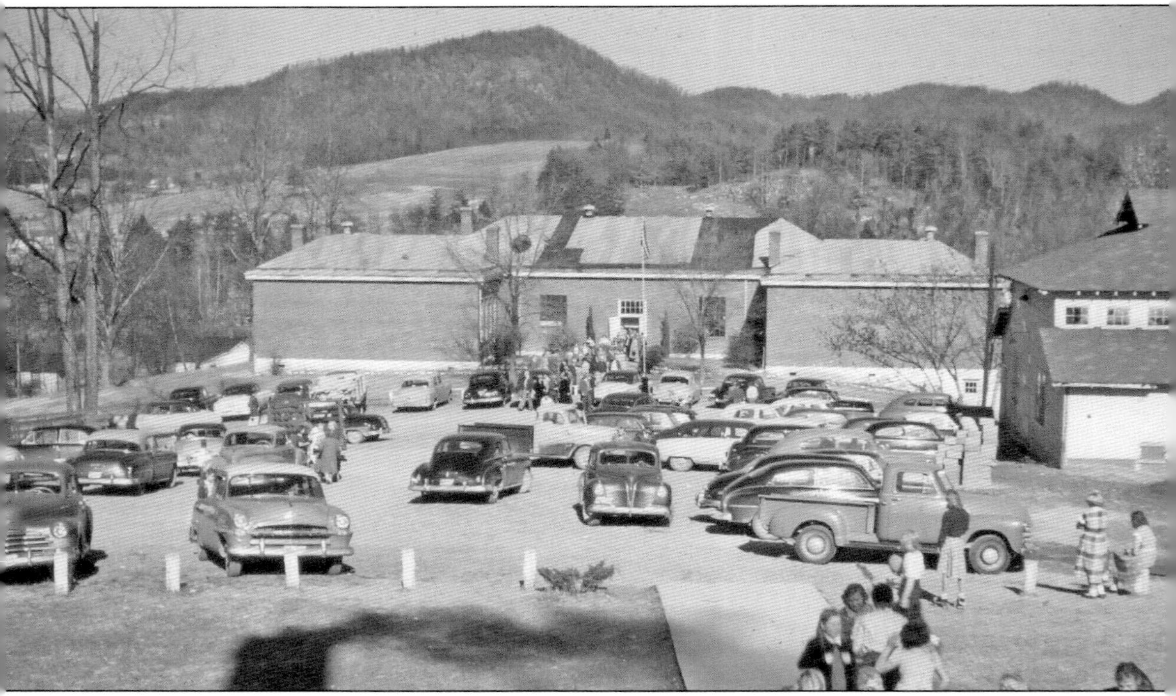

This photograph of Townsend High School was taken sometime between 1948 and 1952. District 15 was the district of the Tuckaleechee Cove area; in 1918, the schools listed were Coker Hill, Steel Pile, Walkers Valley, Red Bank, Frogtown, College Hill, and Sunshine. The College Hill School closed in 1930. (Courtesy of the William Derris Slide Collection, MS.2123. University of Tennessee Libraries, Knoxville, Special Collections.)

Georgia Bradshaw was the daughter of James Homer and Lura Houser Bradshaw. She grew up in lumber camps where her father worked, such as Elkmont, Fish Camp, and Walker's Valley. The family later moved to the Townsend area. Georgia taught school at Red Bank School in Dry Valley in 1938 and 1939, then taught at Townsend from 1939 to 1977. Townsend School was closed in 1977 due to the consolidation of the county high schools. (Courtesy of Dale Bradshaw.)

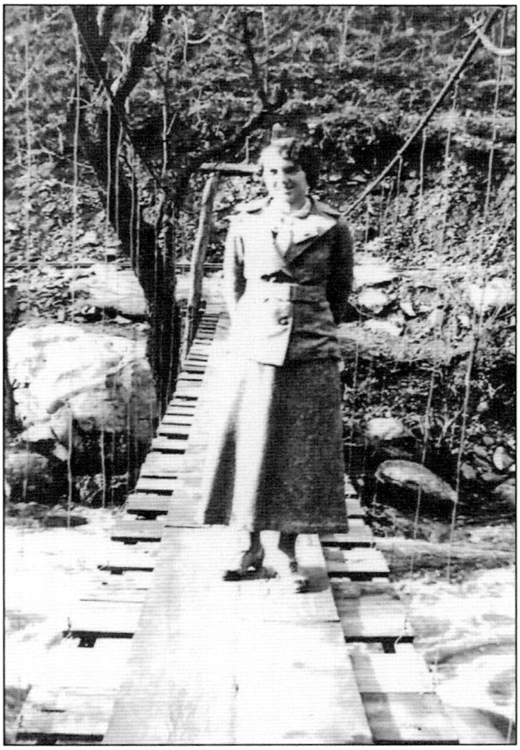

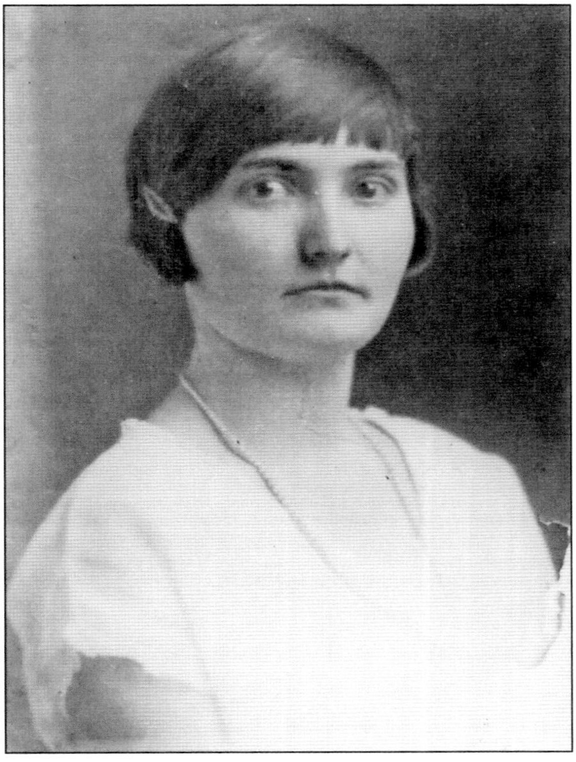

Angie Caylor Myers taught school from 1918 until 1930 at Cades Cove, Coker Hill, Louisville, Red Bank, Caughrons (in Miller's Cove), and Rocky Branch. She then taught at Townsend Elementary from 1930 until 1931. She took time off when her children were small, but when the last one started school, she went back to teaching. Angie retired in 1965. (Courtesy of Marilyn Myers Byrd Photograph Collection.)

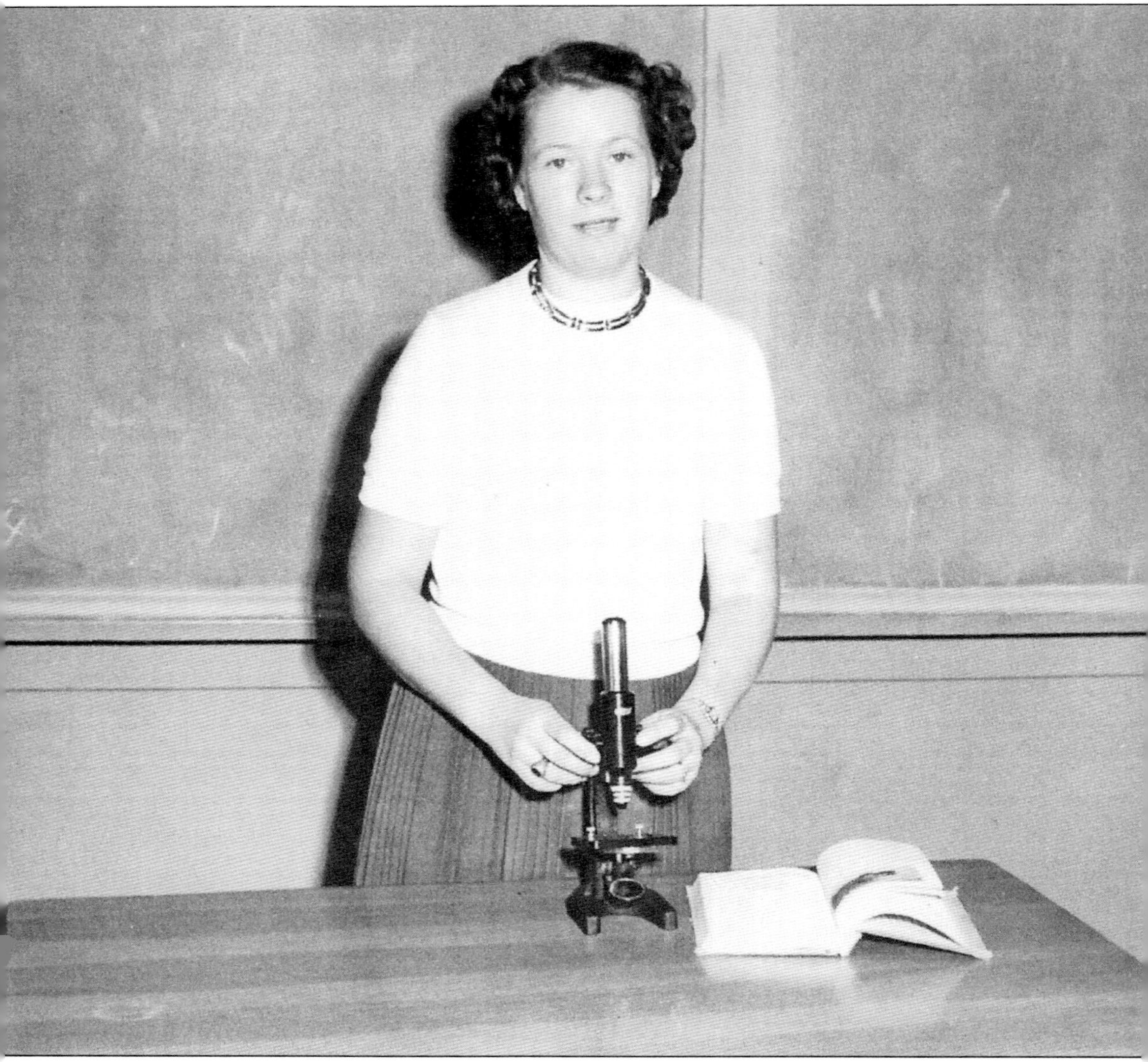

Venenda Tulloch was the science teacher at Townsend High School. There were no science labs but biology and chemistry were still taught. (Townsend Alumni Photograph Collection.)

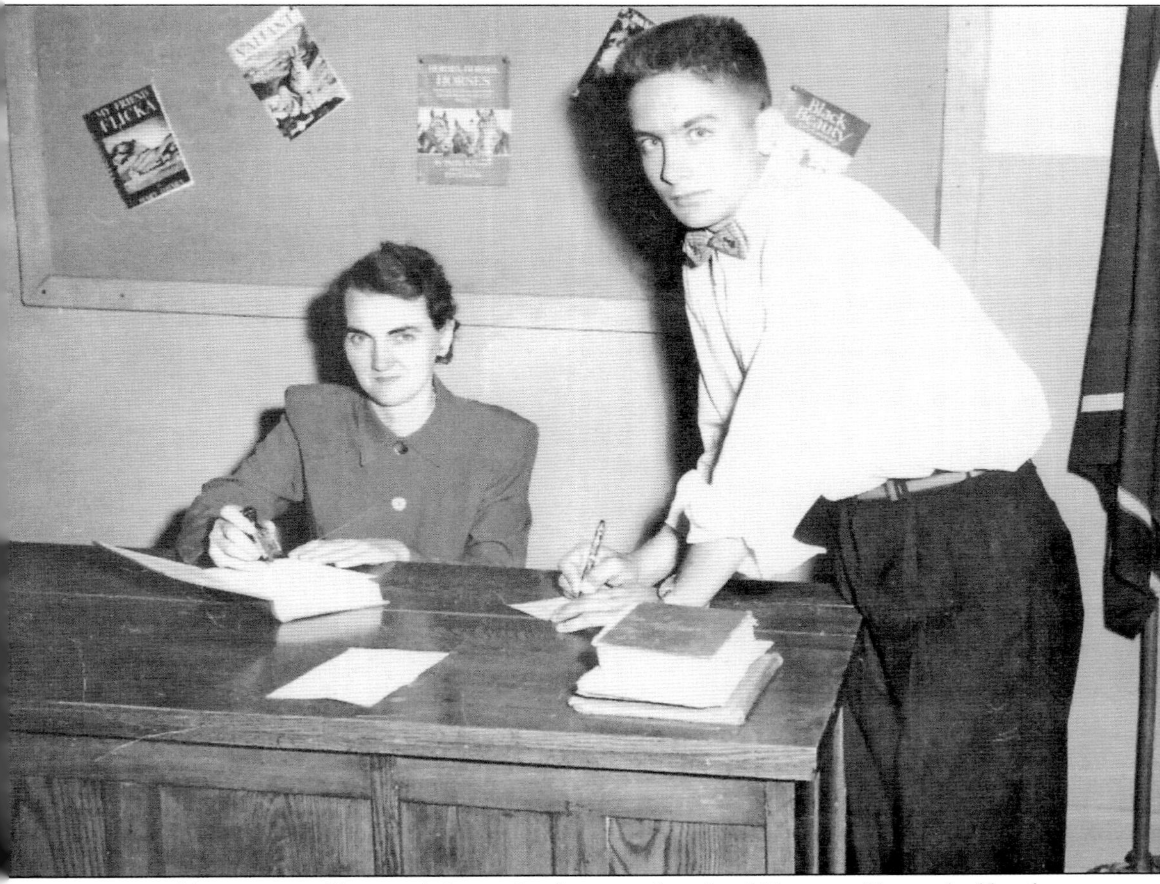

Georgia Bradshaw, pictured here with Jimmy Lindsey, was the school librarian. She worked hard to provide a good library for the students. (Courtesy of Townsend Alumni Photograph Collection.)

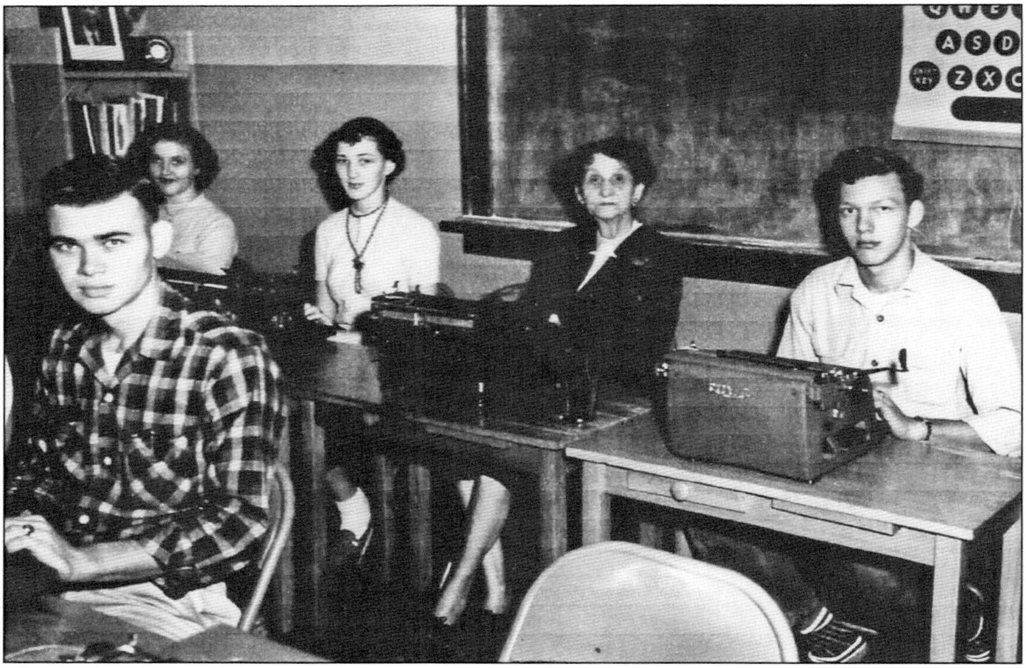

This is the 1954–1955 typing class of Eleanor Adams, a skilled business teacher. Pictured are, from left to right, Durant Tipton (front), Wilma Fancher, Betty Ann Lawson, and Glenn Teaster. (Courtesy of Townsend Alumni Photograph Collection.)

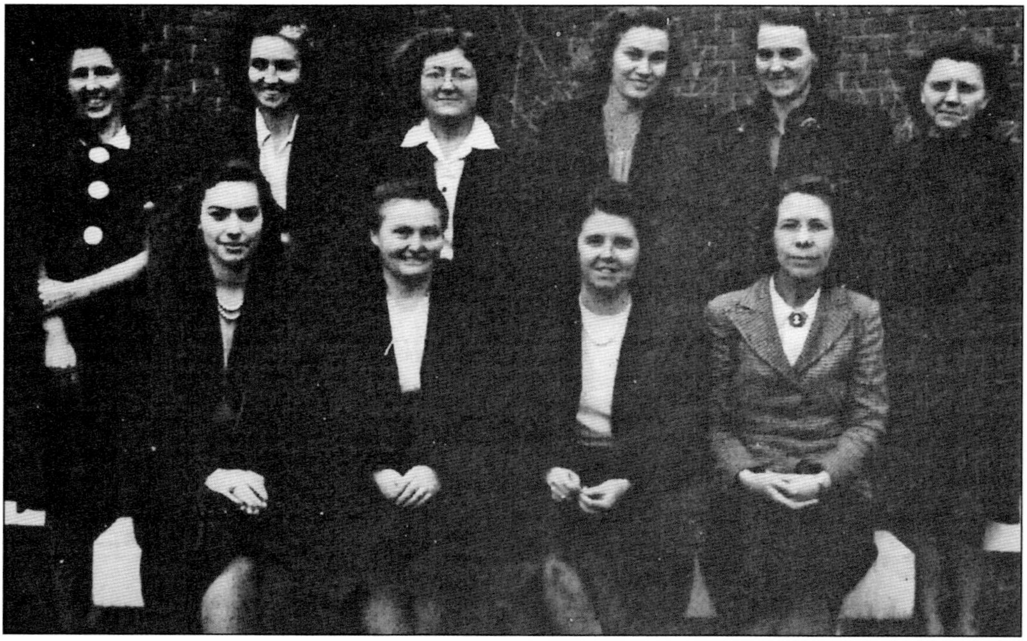

These Townsend School teachers are, from left to right, (first row) Barbara McNiell, Mrs. Lee Myers, Mrs. J. Fred Sentell, and Nina Abbott; (second row) Mrs. Earl Walker, Mrs. Burns, Mrs. L. Reagan Leland, Mrs. Mike H. Davis, Miss George Bradshaw, and Mrs. Charlie Burns. (Courtesy of McNiell Family Collection.)

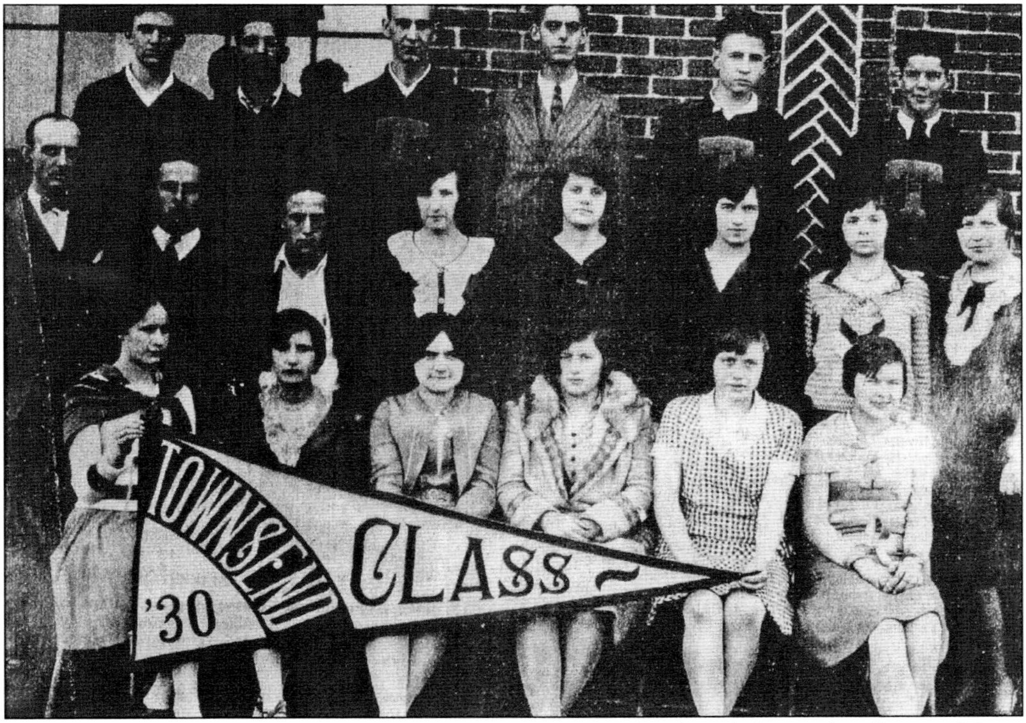

This image of the Townsend graduating class of 1930 includes, from left to right, (first row) Lura Gregory, Irene Keeler, Ocey Myers, Flo Gregory, Ruth Rudd, and Ruth Henry; (second row) Professor Johnson, Henry Gregory, Ollie Gregory, Grace Dunn, Mae Lawson, Mildred Linginfelter, Alberta LeQuire, and Mrs. Johnson; (third row) Ray Patty, Roy Gregory, Wade Tipton, Wade Caylor, Frantz Gregory, and Labe Gregory. (Courtesy of Townsend Alumni Photograph Collection.)

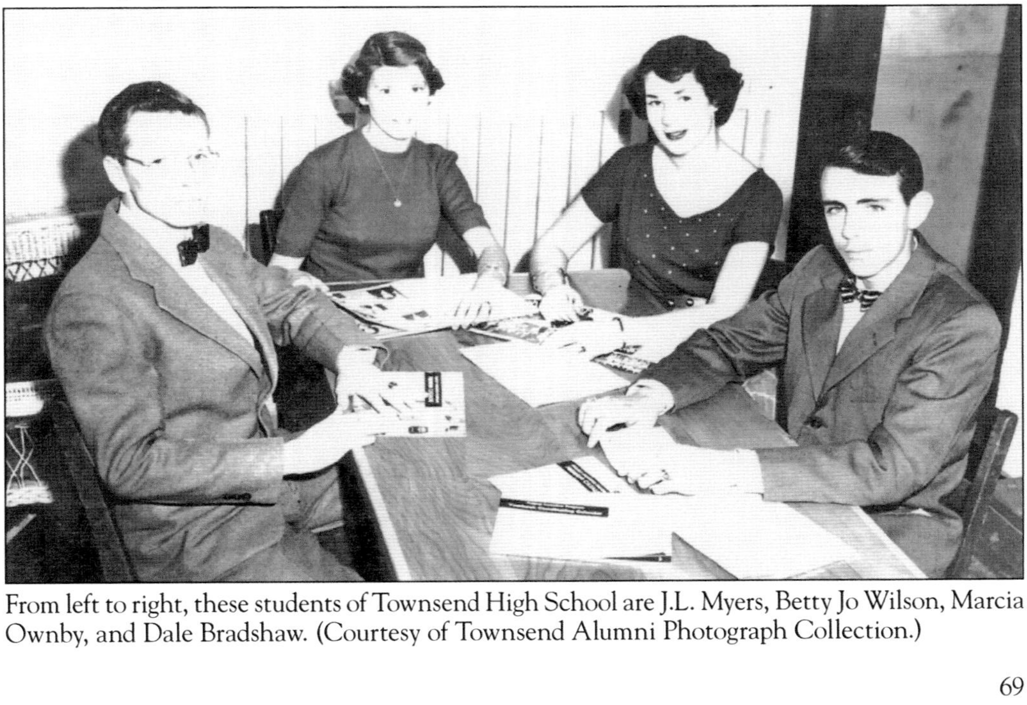

From left to right, these students of Townsend High School are J.L. Myers, Betty Jo Wilson, Marcia Ownby, and Dale Bradshaw. (Courtesy of Townsend Alumni Photograph Collection.)

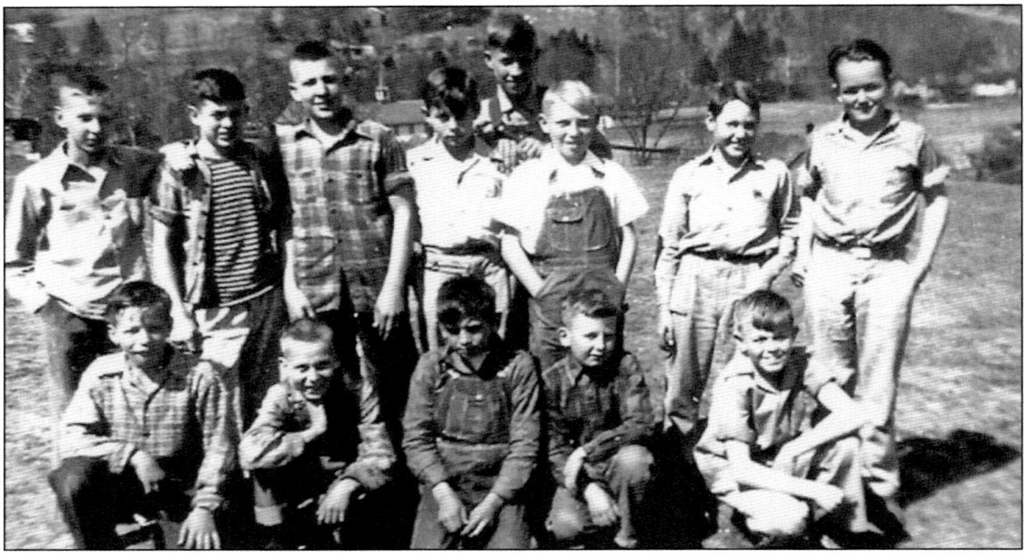

This is Barbara McNiell's 1945–1946 sixth-grade class at Townsend. Students' textbooks were purchased at the J&K Store in nearby Maryville. When a student finished the year with their books, they could sell them to other students. (Courtesy of McNiell Family Collection.)

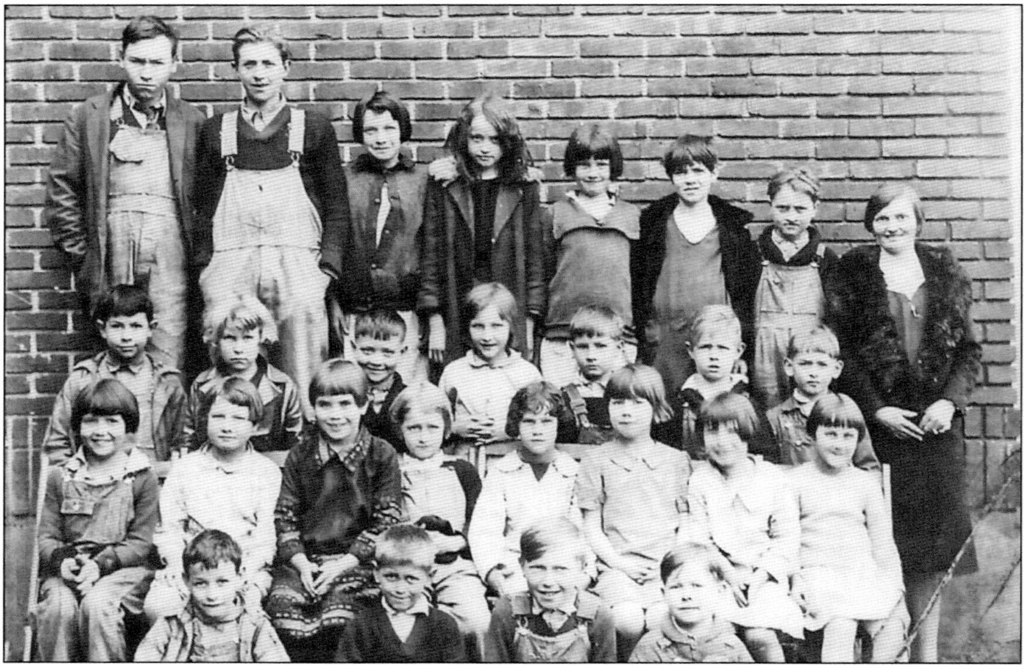

These students attended Townsend School during the 1930–1931 school year. From left to right are (fourth row) Charlie Moore, Jack Myers, Mary Frye, Geraldene Click, Cordia Frye, ? Young, Benny Oliver, and teacher Angie Myers teacher; (third row) Douglas Headrick, Helen Webb Grant, Anderson Jackson, ? Lane, Ray Burns, Glenn Adams, and Don Dorsey; (second row), ? Crye, Mabel Lail, Dorothy Shuler, Rose Reagan,? Hannah, Dorothy Downs, Reba Webb, and Juanita Davis; (first row) George Tipton, Elmer Gibson, Charlie Myers, and ? Frye. (Courtesy of Marilyn Myers Byrd Photograph Collection.)

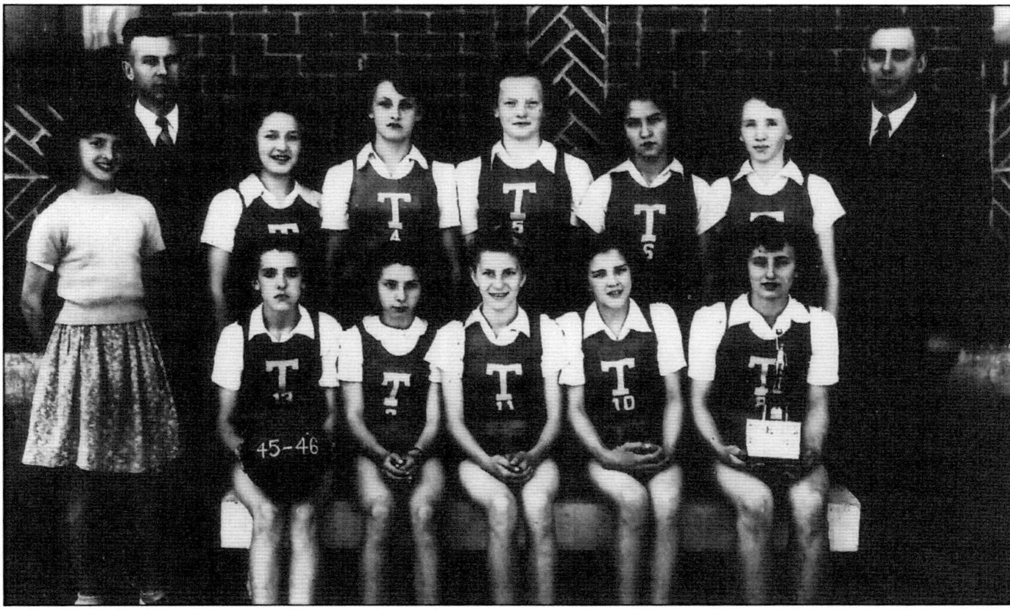

The 1945–1946 Townsend girls' basketball team played half-court basketball with three guards and three forwards on each end. Townsend girls produced some championship teams as well as many talented players. Seating in the Townsend gymnasium was limited at the games because of the overwhelming support of the community. Even though the old Townsend gym had seen its better days, the enthusiasm of the players and support from the fans kept the orange-and-white spirit alive. The fans came out and supported the Townsend team no matter the opponent. (Courtesy of Townsend Alumni Collection.)

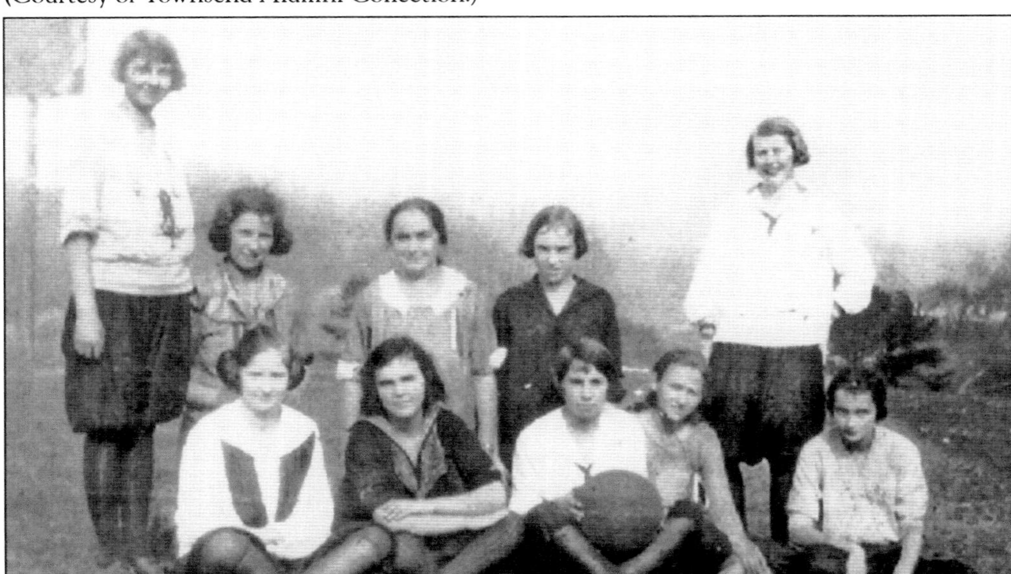

In 1923, the Townsend High School girls' basketball team played on a dirt court (the basket can be seen to the left). Stella Myers is holding the basketball. Girls played half-court basketball with three guards on one end, two forwards, and a center on the opposite end. (Courtesy of McNiell Family Collection.)

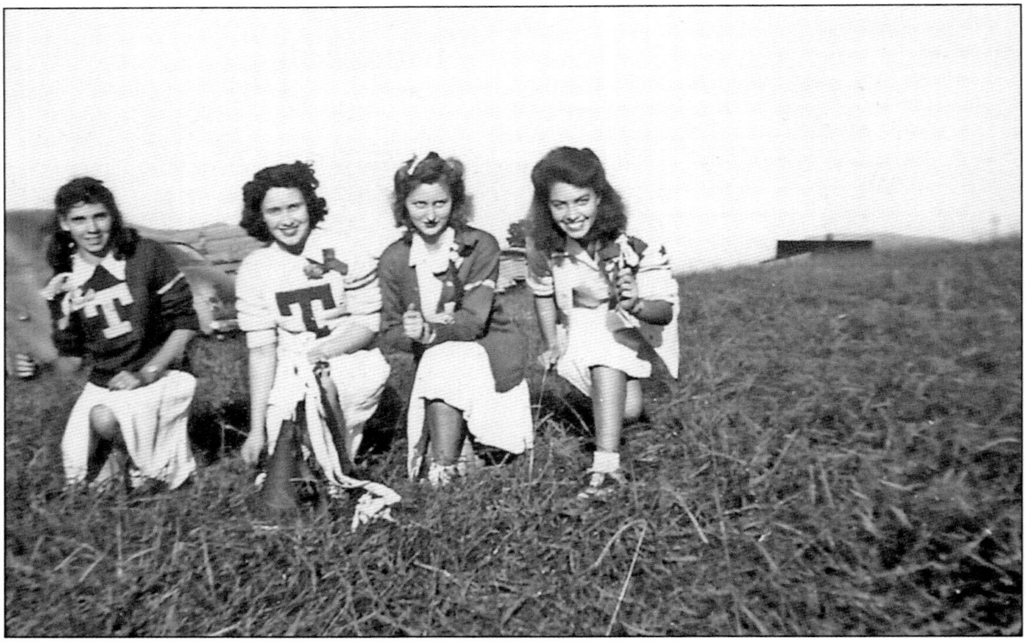

From left to right, these Townsend High School cheerleaders are Mary Ruth King, Johnnie Hatcher, Evelyn Wear, and Barbara McNiell. (Courtesy of McNiell Family Photograph Collection.)

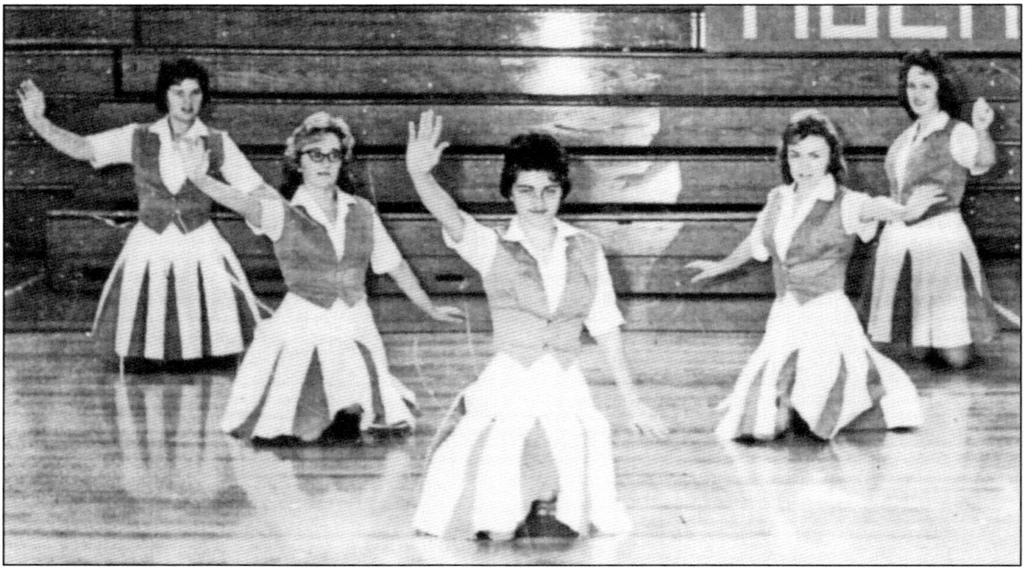

Pictured here from left to right are Townsend High School cheerleaders Carolyn Neff, Sugie Cromwell, captain Donna McClanahan, Linda Grant, and Donna Hurst. (Courtesy of Townsend Alumni Photograph Collection.)

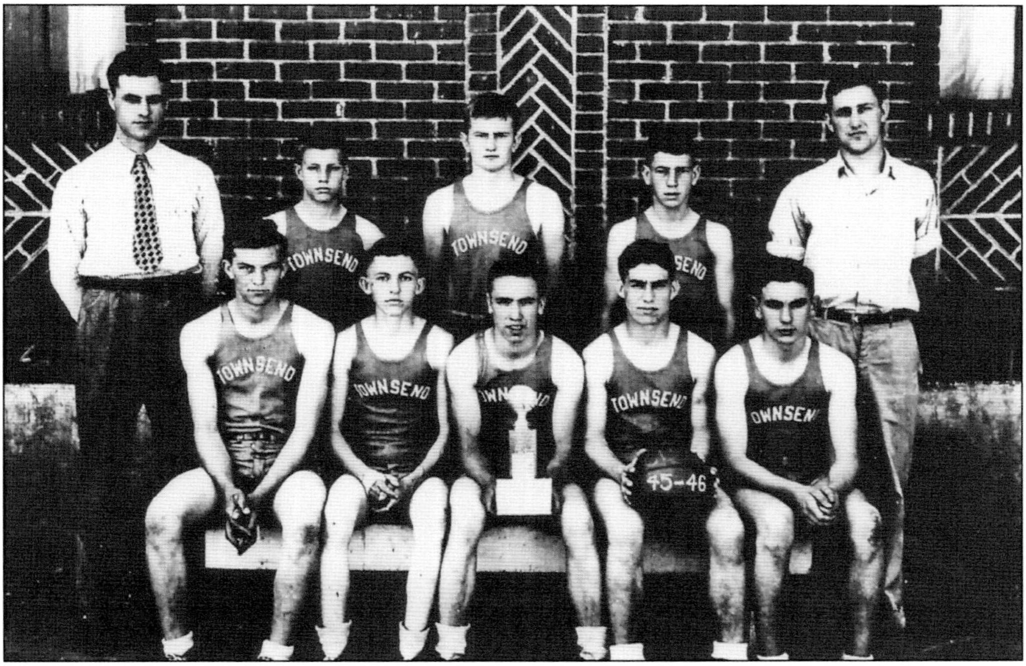

Pictured here is the 1945–1946 Townsend men's basketball team. From left to right are (first row) Willie Brickey, Sammy Adams, Ralph Myers, Joe McNeill, and C.H. Bradshaw; (second row) coach Mike Davis, Dale Carnes, Jack Webb, James Dockery and manager Wayne Myers. (Courtesy of Townsend Alumni Photograph Collection.)

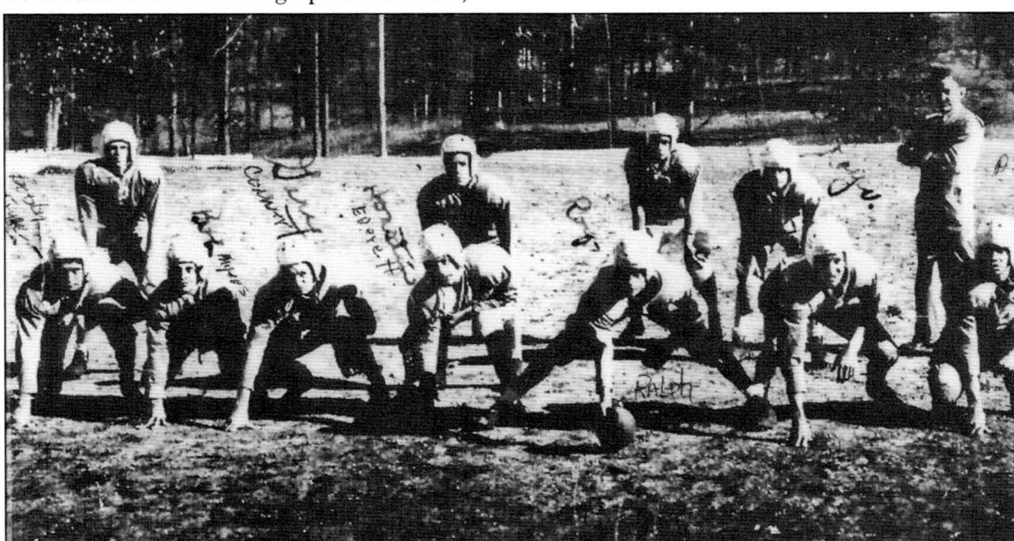

To prepare for the upcoming football season, players went through conditioning. One of the camps was held in nearby Cades Cove. The players camped out nightly and began an early morning of conditioning that usually lasted all day. The players' families provided food and provisions, and cooks went along to prepare food. In later years, camps were held at Laurel Lake in nearby Dry Valley. Laurel Lake had barrack-like buildings for the players to sleep in. (Courtesy of Townsend Alumni Photograph Collection.)

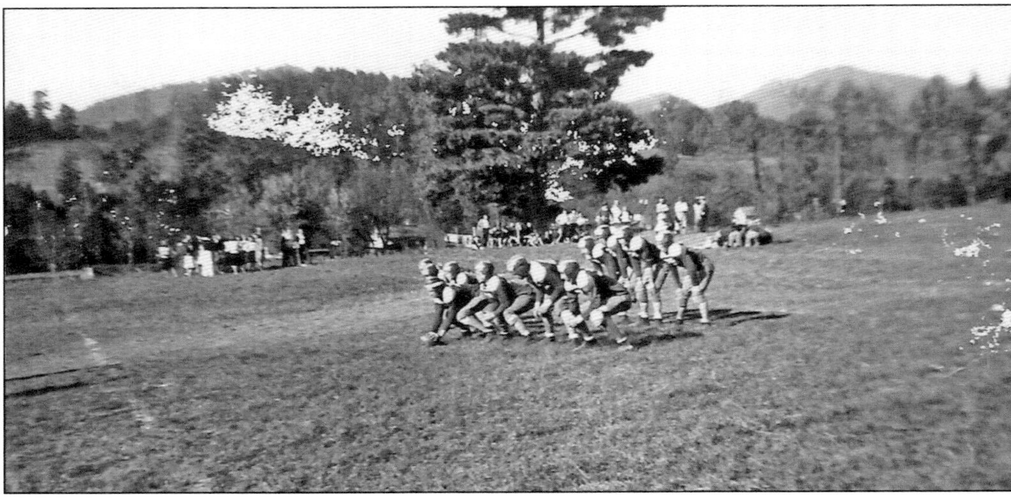

Athletic teams were a very big part of both Townsend High School and the community. When the first football team was organized in 1926, players wore purple and gold and were known as the Lumberjacks. In order to have uniforms, the first team had to furnish their own shoulder pads and shoes; the principal at the time purchased 12 pairs of pants. Helmets and headgear were not required then. The practice field was located at what is now known as Valley View Lodge. The colors of the team changed to orange and white, and the team became known as the Tigers. (Courtesy of Townsend Alumni Photograph Collection.)

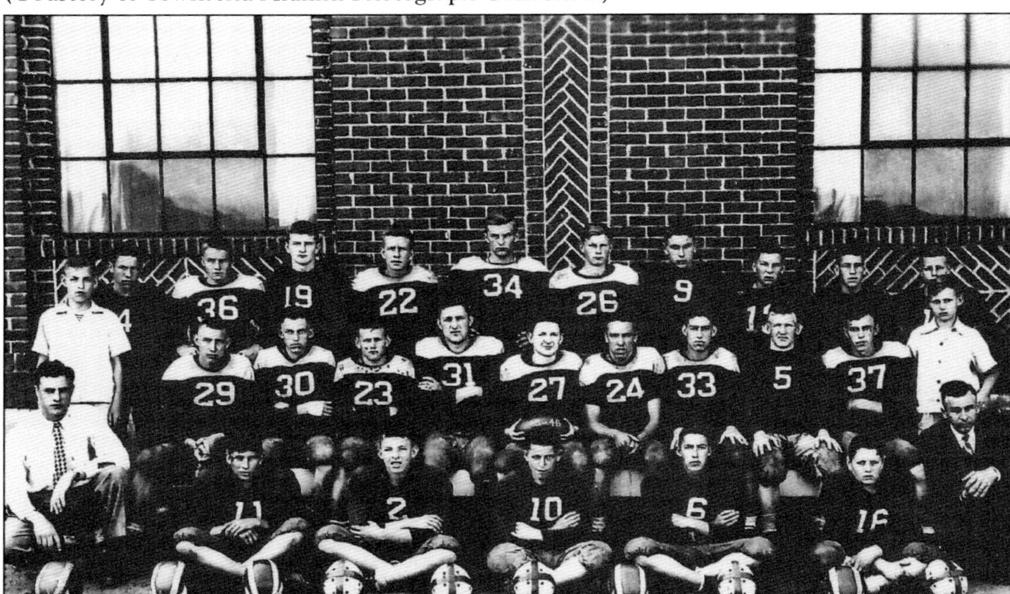

Pictured in this image of the 1945–1946 Townsend football team are, from left to right, (first row) coach Mike Davis coach, Junior Johnson, Ralph Myers, Bill Effler, Donald Foster, Wayne Myers, and principal J. Fred Sentell; (second row) manager Bill Scott, Carl Myers, Sammy Adams, Rex Woodby, Wayne Myers, Bill Franklin, Ralph Myers, Joe McNiell, Junior Fagg, Otis Tipton, and manager Van Myers; (third row) John Spangler, Herbert Woodby, Jack Webb, C.P. Myers, Willie Brickey, Avery Lail, C.H. Bradshaw, Kenneth Lail, James Dockery, and Dale Carnes. (Courtesy of Townsend Alumni Photograph Collection.)

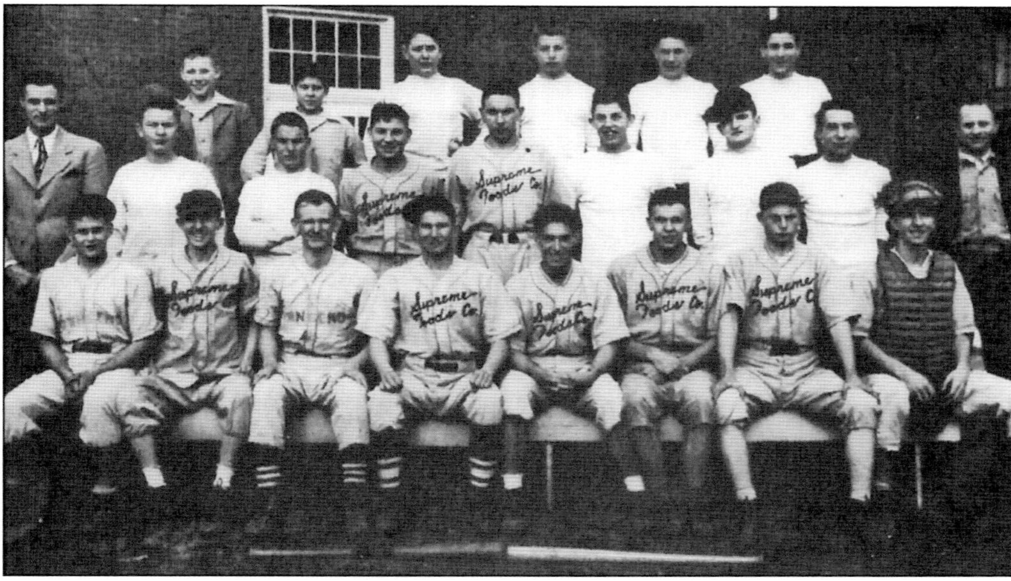

Townsend baseball games were held on Saturday nights with teams from Walland, Alcoa, and Maryville. Many fans in Blount County came to watch the games. (Courtesy of Townsend Alumni Photograph Collection.)

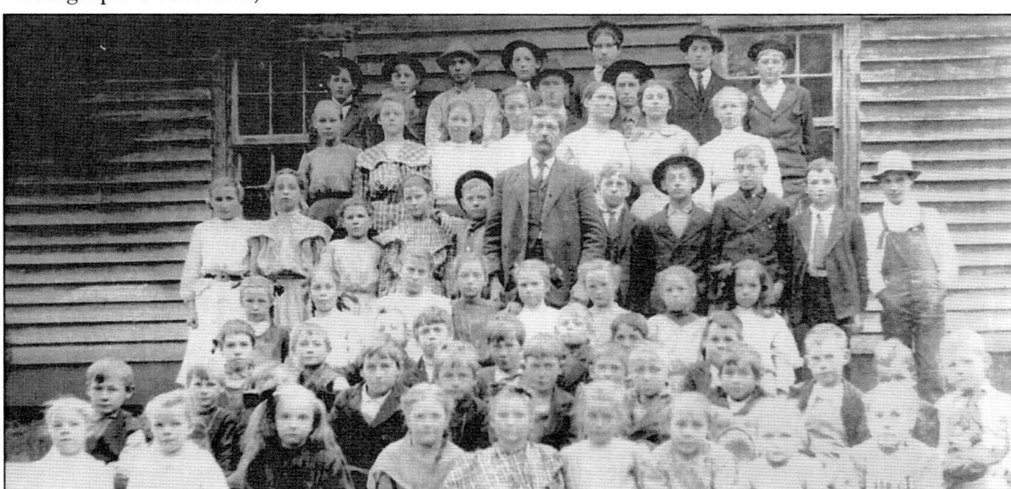

These are the students of Coker Hill School during the 1910–1911 school year. From left to right are (first row) Nancy Brickey, Lela Abbott, Ruby Gilbert, Bessie Law, Zola Burns, Rebecca Dunn, Lydia Adams, Alice Dunn, and Nina Abbott; (second row) Geoffrey Adams, Charlie Adams, unidentified, Dexter Rathbone, two unidentified children, Arthur Cromwell, Ray Rathbone, Oscar Walker, Bryson Dunn, and Harrison Dunn; (third row) John Brickey, three unidentified children, Peter Dunn, and three unidentified children; (fourth row) Maude Brickey, Roxie Jenkins, Martha Russ, Belva Dunn, Florence Brickey, Ida Abbott, Susie Brickey, and Carrie Gilbert; (fifth row) unidentified, Lizzie Headrick, Cora Adams, Lydia Burns, unidentified, teacher John Brickey, unidentified, Mathis Dunn, Bell Headrick, and Lincoln Frye; (sixth row) Alice Adams Rudd, Delia Burns, Josie Burns, unidentified, Flora Burns, unidentified, Emma Brickey, and Ed Adams; (seventh row) Clifford Adams, Mell Adams, Oscar Sutton, Mell Brickey, Earnest Rathbone, E. Adams, and Lee Brickey. (Courtesy of Joann Shuler Photograph Collection.)

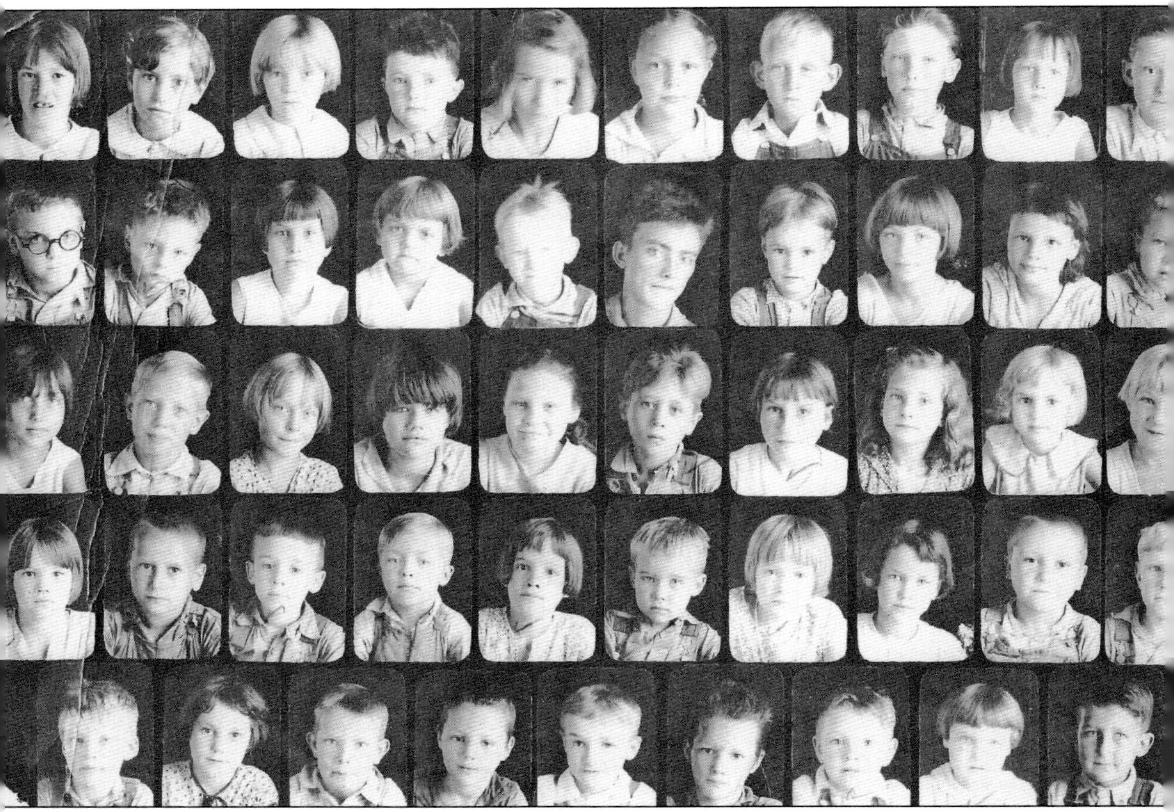

This 1930 group from Coker Hill School was taught by Pearl Caylor. Pictured are, from left to right, (top row) ? Headrick, Elizabeth Compton, Beatrice Brickey, Conard Hembree, Edith Headrick, Dorothy Sullivan, Ray Adams, Eugene Hyman, Johnie Faye Brickey; and unidentified (second row) Clarence Cromwell, ? Ogle, Helen Brewer, Bernice Cromwell, Ray Webb, Arlie Trentham, Carlos Hembree, Josie Brickey, Clara Fay Ogle, and Clyde Emert; (third row) Juanita Brickey, Richard Walker, Ezalee Henry, Nandy Ogle, Ada Brewer, ? Hembree, Hassel Walker, Josie Honeycutt, ? Webb; Mary Sue Dunn (fourth row) ? Headrick, Harold Rathbone, unidentified, and Carl Dunn, Merle Headrick, unidentified, ? Headrick, Minnie Davis, Sherill Rathbone, and Robert Dunn; (fifth row) unidentified, Bell Frye, John Walker, Billie Ogle, Curtis Webb, Earnest Ogle, Kenneth Sullivan, Bell Carroll, and Mayford Brewer. (Courtesy of Paulette Ledbetter.)

Three

Buildings, Homes, and Farms

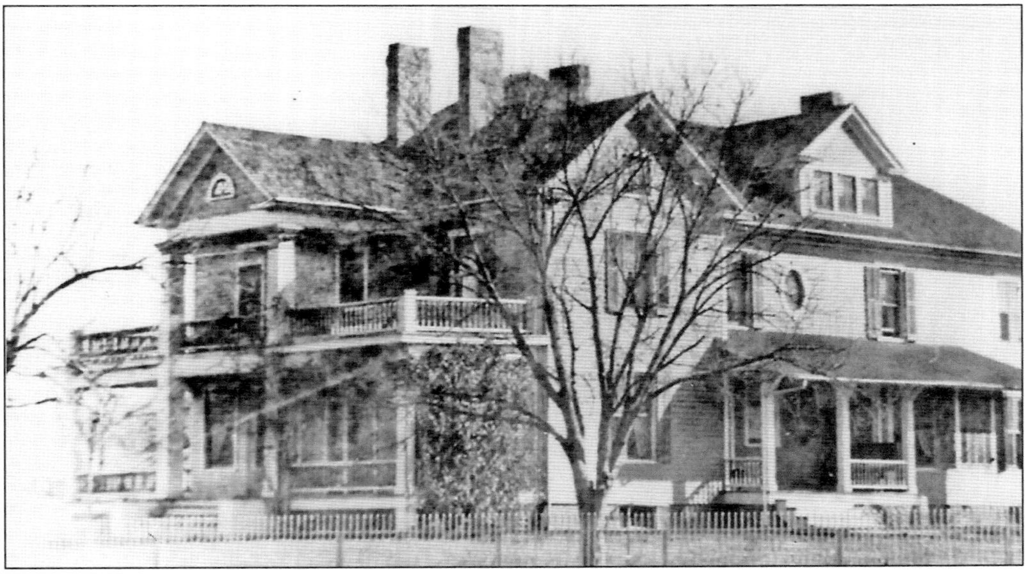

This is the home of W.B. "Wilson" Townsend, president of Little River Railroad. Townsend bought many acres of land in the early 1900s to start a lumber business. This house sat in close proximity to the Little River Lumber Company, so he was able to keep an eye on the operations. W.B. Townsend was a very generous man to the community and to the schools. Though he still had other responsibilities in his former states of Pennsylvania and Kentucky that required him to spend lots of time there, his permanent base of operations was in Townsend. (Courtesy of Great Smoky Mountains National Park.)

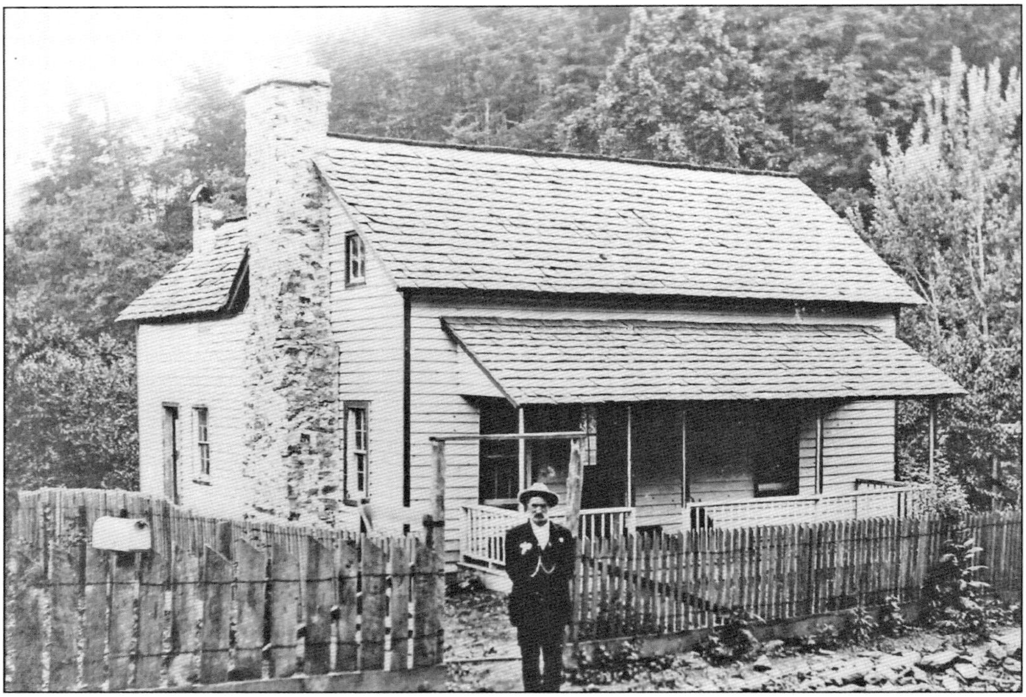

William Myers's built his house on Webb Road at Carr's Creek Road. Myers was born on April 8, 1846, and married Mary E. Lowe on December 6, 1866. Their children were William Houston, Sarah, Texana, Henry T., Nathan R., Peter, Doc, Carrie, Dora, Elizabeth, and Elmer. (Courtesy of Charlie Myers.)

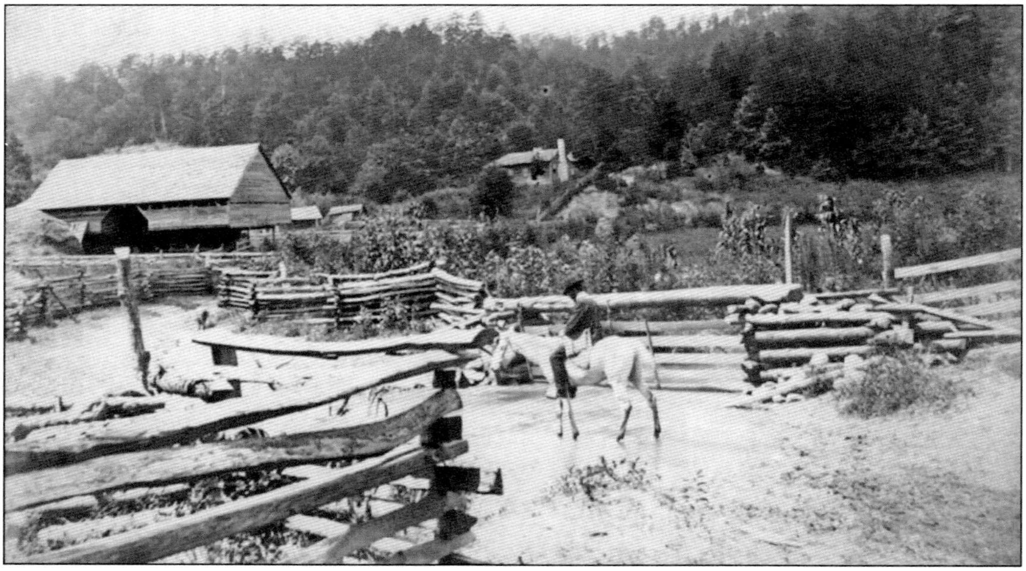

The Com Caughron farm in Tuckaleechee Cove is pictured here in August 1886. William Cox Cochran, who had traveled into the area, took some wonderful photographs. The barn in the picture is a cantilever barn, common for the area. (Photograph by W.C. Cochran, courtesy of the University of Tennessee Libraries, Knoxville, Special Collections.)

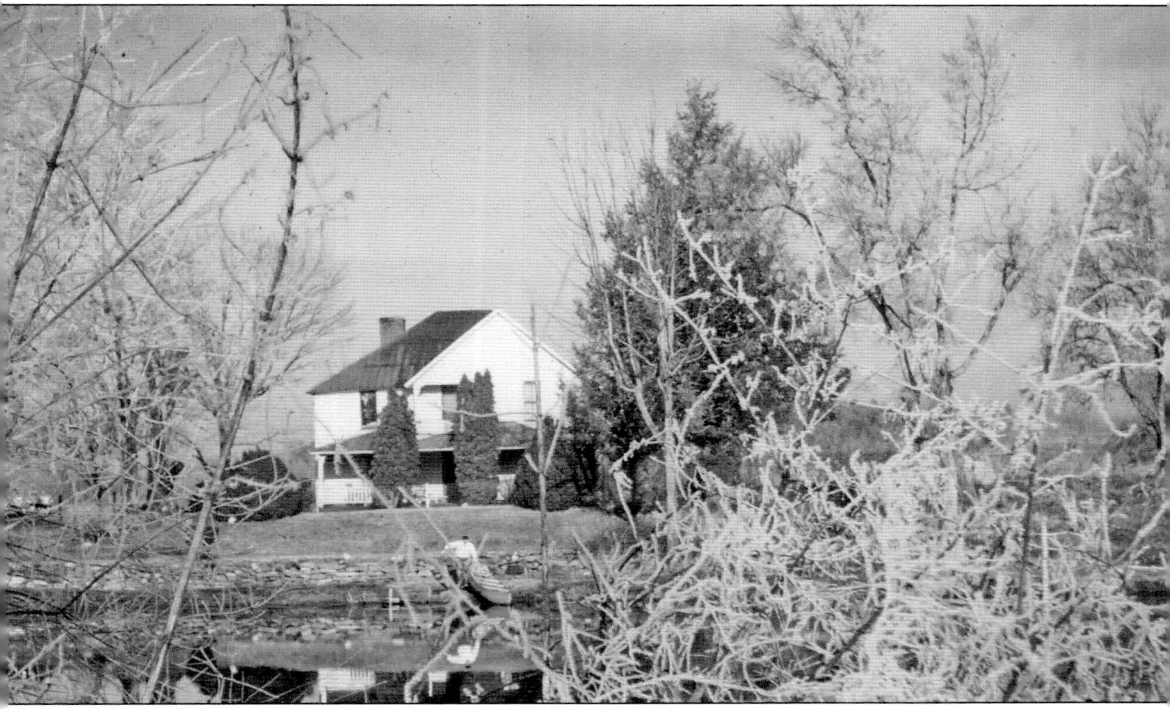

William Robinson "Bill" and Mary Alice Rambo Dunn lived in this house built by the Little River Lumber Company near the dam. This dam was used to produce electricity for the lumber company. The electric company was known as the Townsend Light & Mill Company. (Courtesy of the William Derris Slide Collection, MS.2123. University of Tennessee Libraries, Knoxville, Special Collections.)

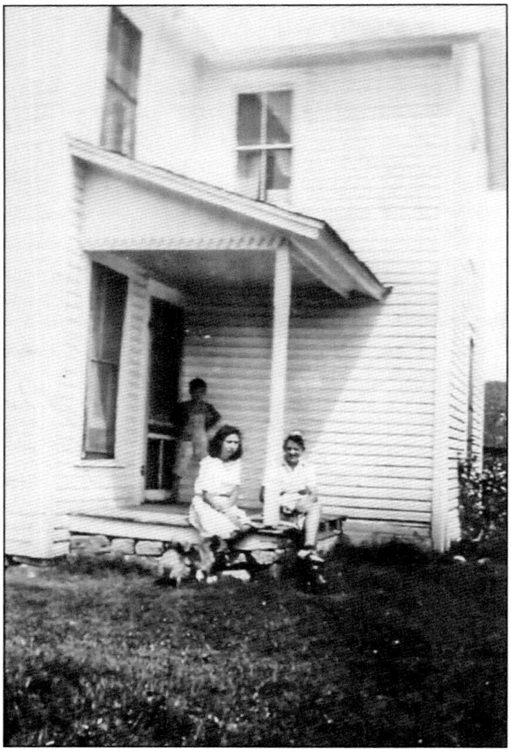

Golman Myers bought this house around 1940 from Bill Cotter. Cotter was married to the daughter of the original owner, Joseph "Joe" Walker, who had built the home around 1900. Golman purchased 85 acres facing Little River. The Myers family was known for their cattle farming. They raised Hereford cattle, although Golman would mix his Durham bull with his Hereford cow, and the resulting cattle would produce better milk and better beef. Though Golman Myers's family moved into the house around 1940, he remained in Cades Cove taking care of cattle. He passed away while he was still in the cove. (Courtesy of Bernard Myers.)

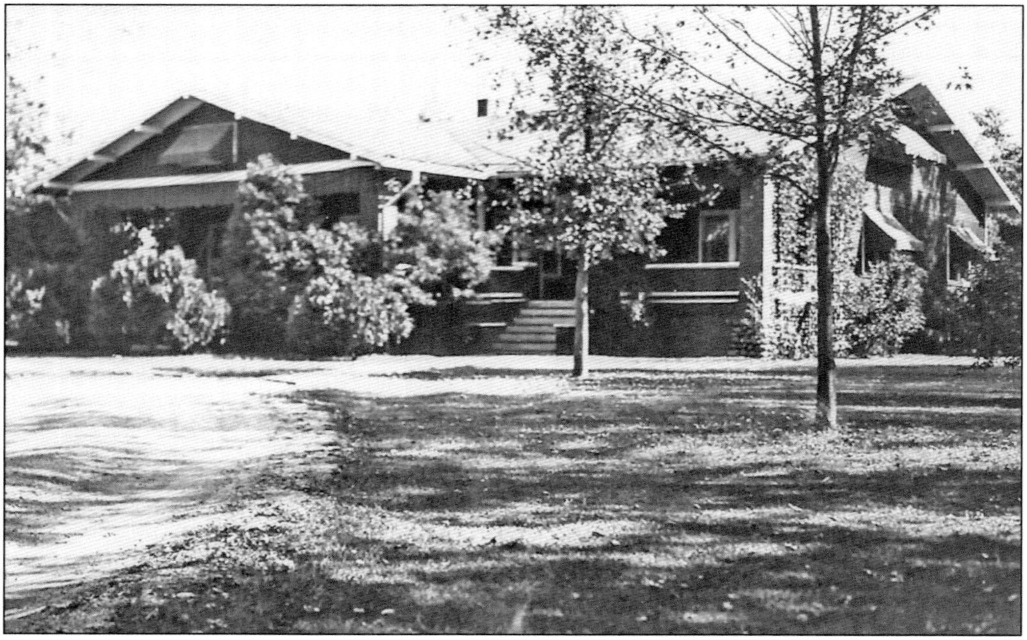

The D.H. "Doc" Tipton house was located near the site of the Little River Lumber Company. Doc served for many years as the superintendent for Little River Railroad & Lumber Company. After the closure of the lumber company in 1940, he continued to be a business leader in the community. (Courtesy of Missy Tipton Green.)

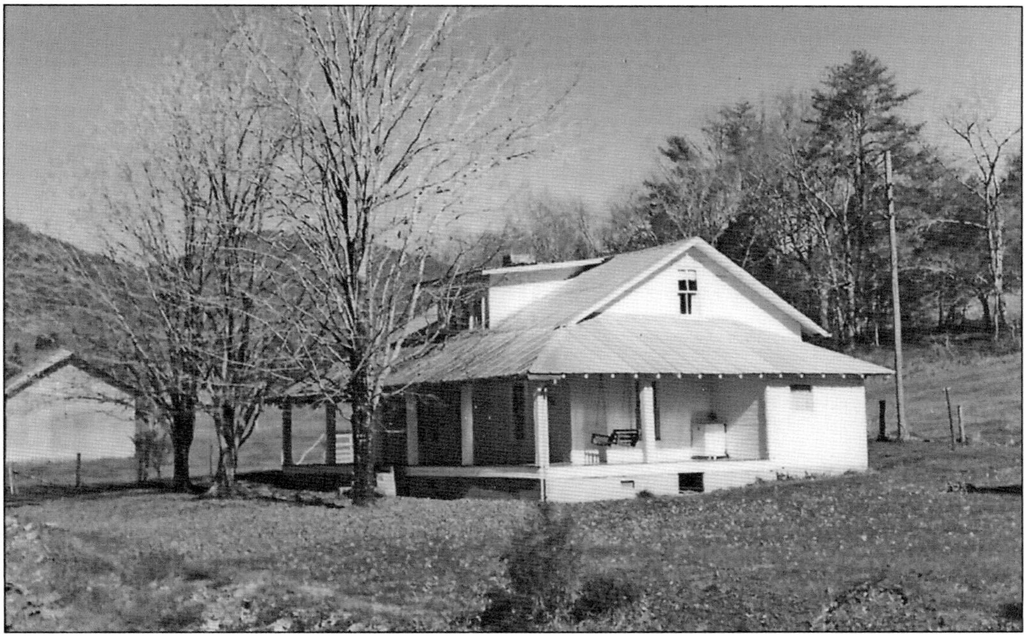

This house was built for George and Mary Caylor on Cedar Creek Road by Bob Dunn and Perry Abbott. Most houses had feather beds covered with thick, handmade quilts; some had straw tick mattresses that had to be refilled each year. Kerosene lamps were used to light the house. The rolling stores would also come through the neighborhood, exciting the children. Mary Caylor traded her eggs and tomatoes to the storekeeper for staple goods. (Courtesy of Marilyn Myers Byrd Photograph Collection.)

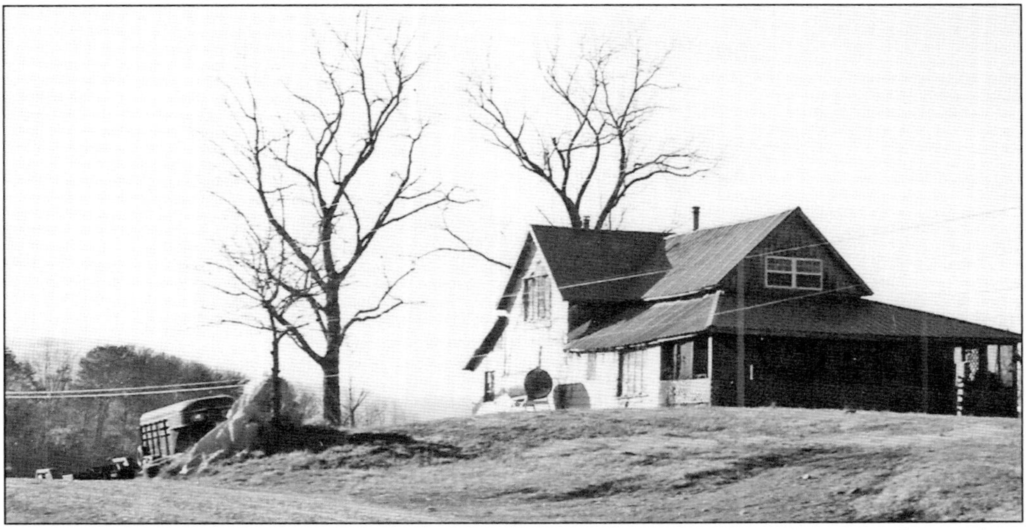

Fonzie and Annie Jane Headrick Bryant's home place was in Dry Valley. Their son John Andrew Bryant married Ruthie Angeline Ledbetter on September 15, 1912. After the death of Fonzie, John and Ruthie purchased the house and moved in. John was a mail carrier, and he picked up the mail in Townsend and Walland, took it to Maryville, then loaded up the mail from Maryville and delivered it to Walland and Townsend. Twice a week, he delivered mail to the Cades Cove area. (Courtesy of Johnnie Bryant Sparks.)

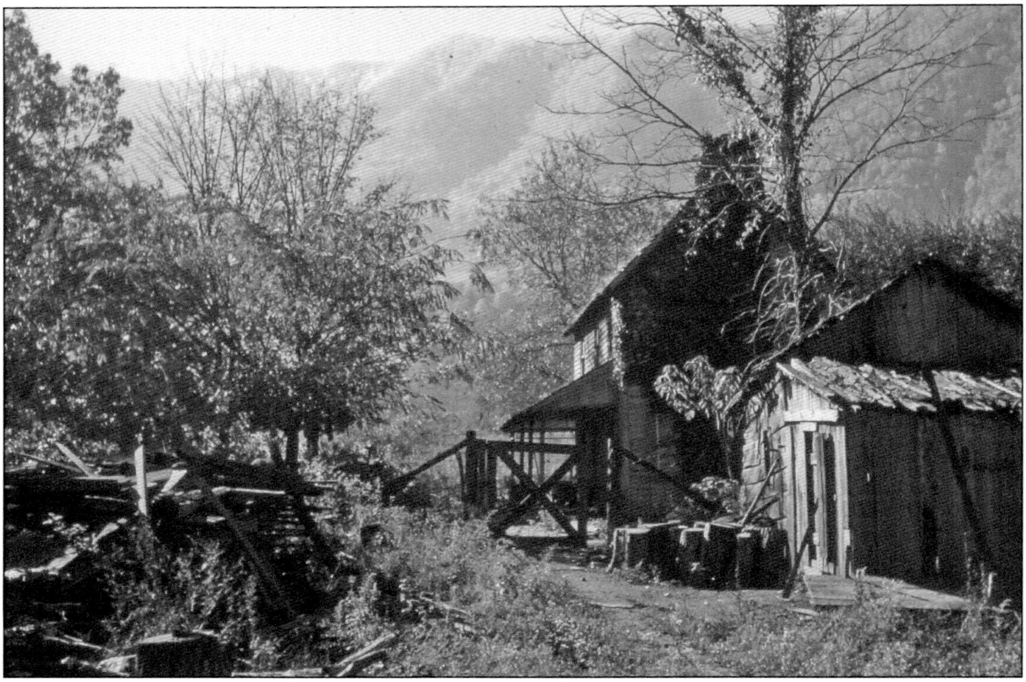

This image shows the Doshie Davis house in the Dry Valley area. The women of this era had to make their own soap and washing detergent. Wood ashes would be saved for the making of the lye; it was then mixed with grease and boiled to make soap. (Courtesy of the William Derris Slide Collection, MS.2123. University of Tennessee Libraries, Knoxville, Special Collections.)

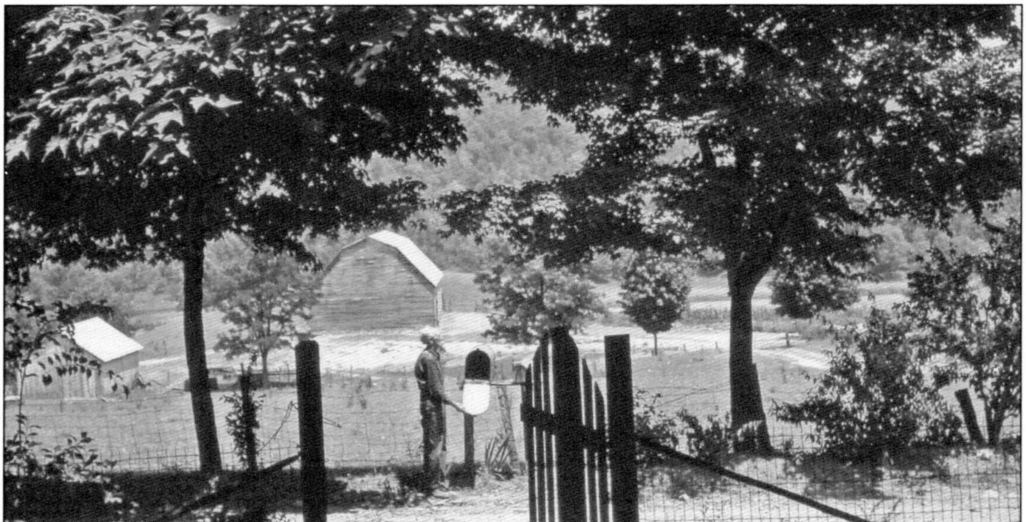

Here, John Dunn checks his mailbox for the daily mail run. The Dunn farm was located in the Dry Valley area on Schoolhouse Gap Road. Electricity did not arrive in the area until about 1900; until that time, kerosene coal oil was used in the lanterns for light. Stoves were heated by wood, which was cut and stacked by family members. From 1930 to around 1950, warm morning heaters were used in the front rooms, and bedrooms were always cold. (Courtesy of the William Derris Slide Collection, MS.2123. University of Tennessee Libraries, Knoxville, Special Collections.)

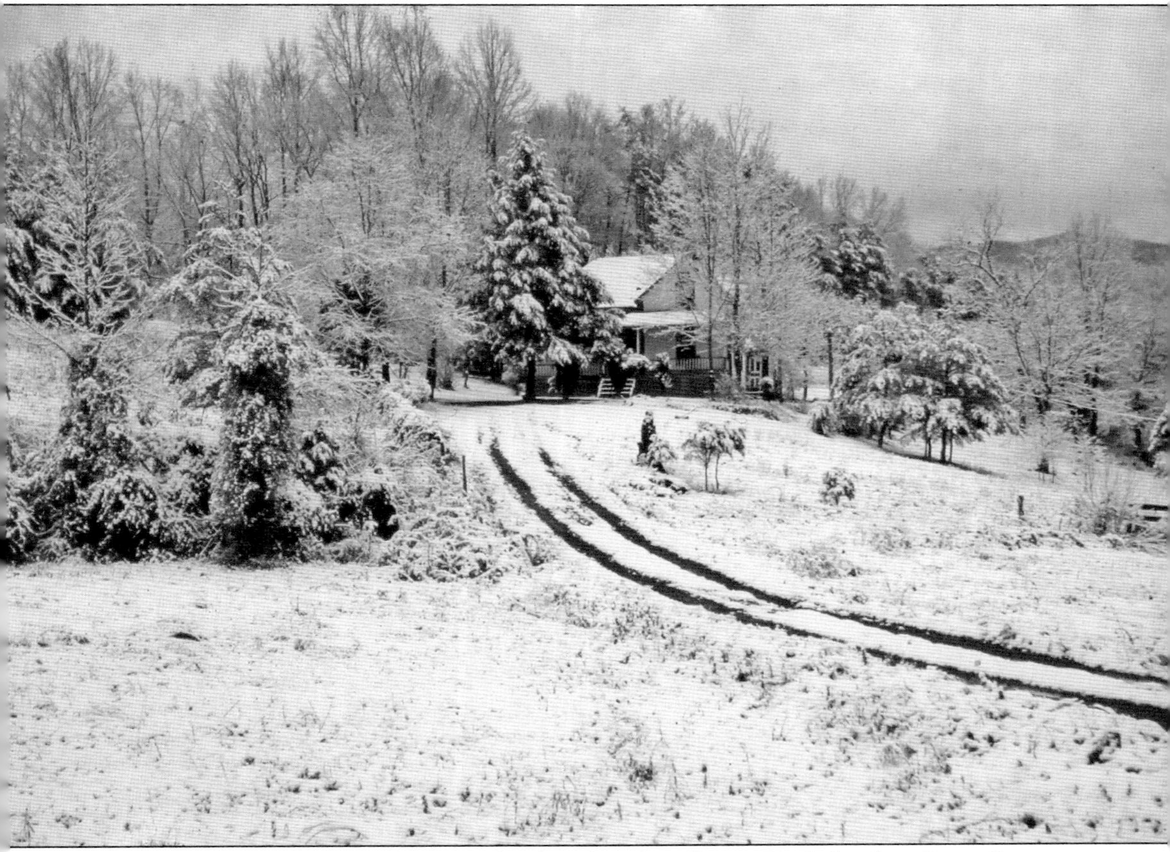

Pictured here is the home of Ruth (McMahan) and Earl Walker in Tuckaleechee Cove. During the late 1930s and early 1940s, Townsend residents read of the warnings of World War II. Life was rough during that time; goods were rationed, so each month families received a book of ration coupons. Many of Townsend's men were drafted into war, and the women and girls took on what were traditionally men's jobs. (Courtesy of the William Derris Slide Collection, MS.2123. University of Tennessee Libraries, Knoxville, Special Collections.)

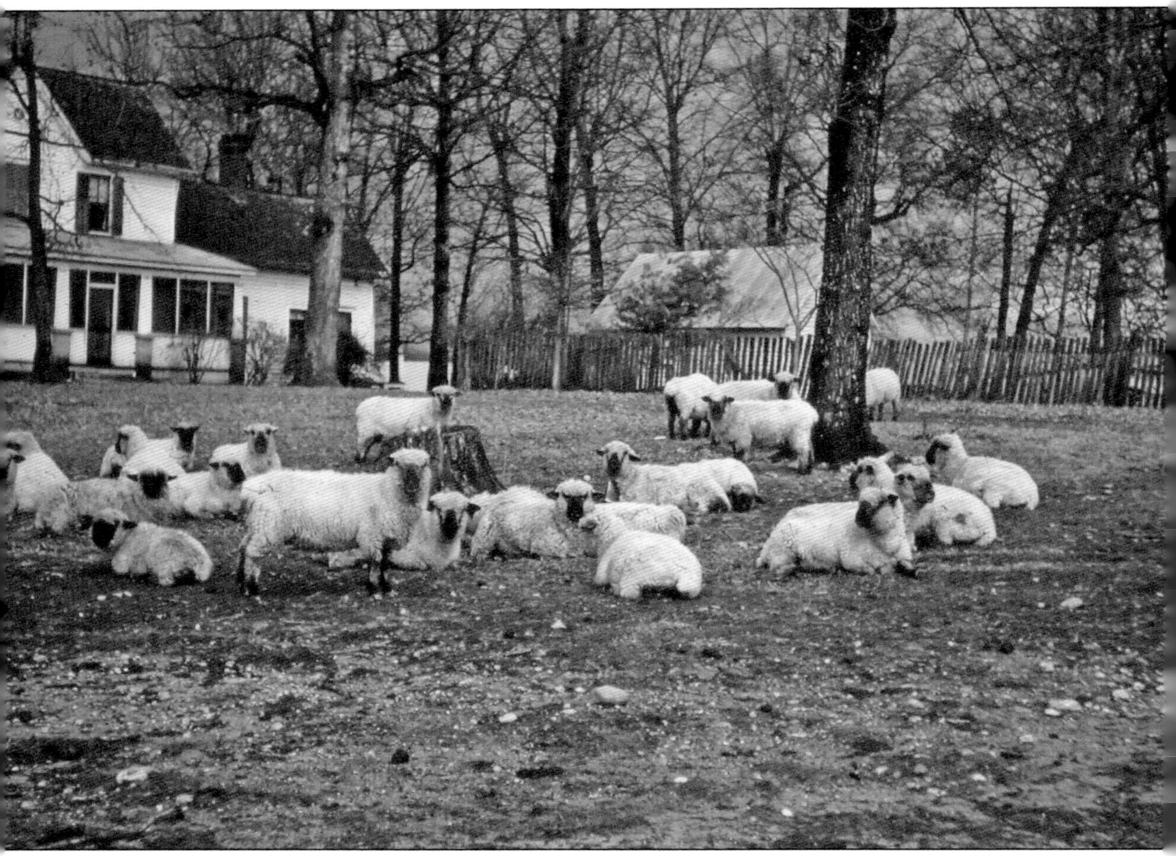

Sheep, like these at Gamble Farm in Dry Valley, were used by many for making wool items. They were sheared, and the wool was cleaned and carded before being spun with a spinning wheel. It was then be used to make coverlets or clothing for the family. (Courtesy of the William Derris Slide Collection, MS.2123. University of Tennessee Libraries, Knoxville, Special Collections.)

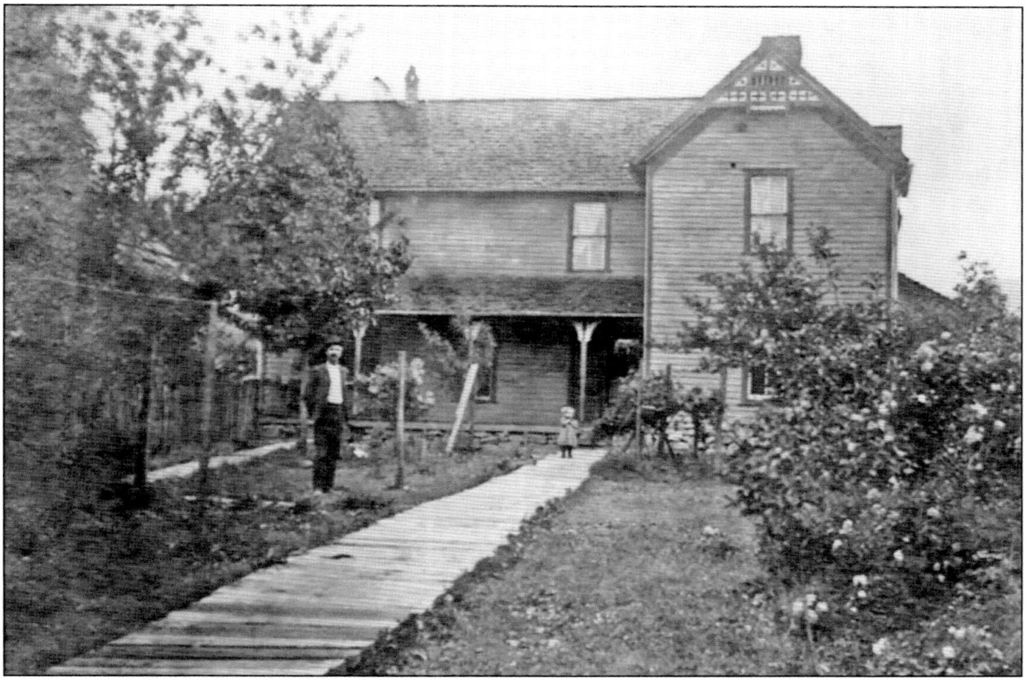

Seen here in 1908 is the home of Joseph "Joe" and Laura Feezell Myers, with Joe and daughter Stella standing in front. Joe inherited this land from his father, William Myers. The wooden boardwalk that Stella stands on leads out to the well house. The foundation of the house was made of native stones, and the structure had two chimneys. There was gingerbread molding on the gables, and the home also had a wraparound porch. (Courtesy of McNiell Family Collection.)

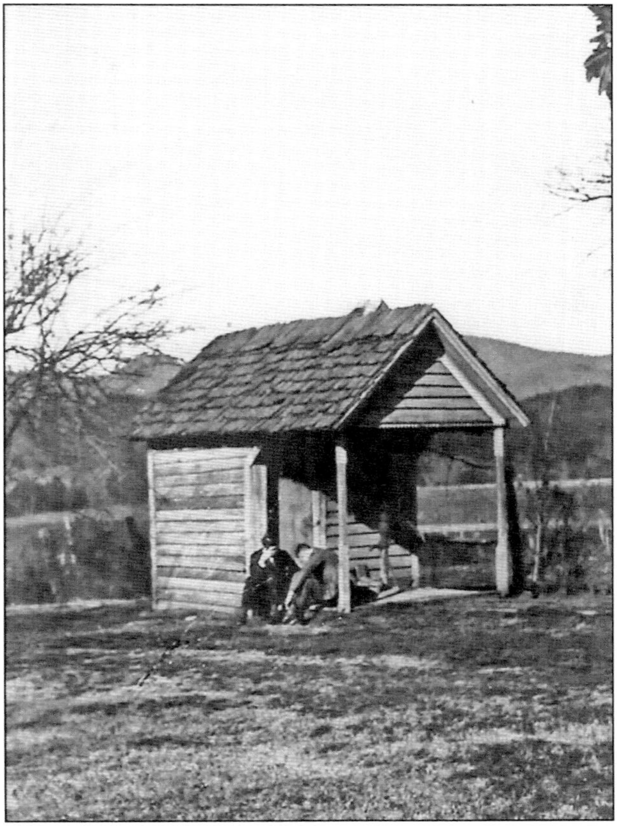

Leonard Myers sits at the well house located behind the Joseph Myers home. The small building was constructed to cover the well and pump from the weather and was also used to store items. (Courtesy of Marilyn Myers Byrd Photograph Collection.)

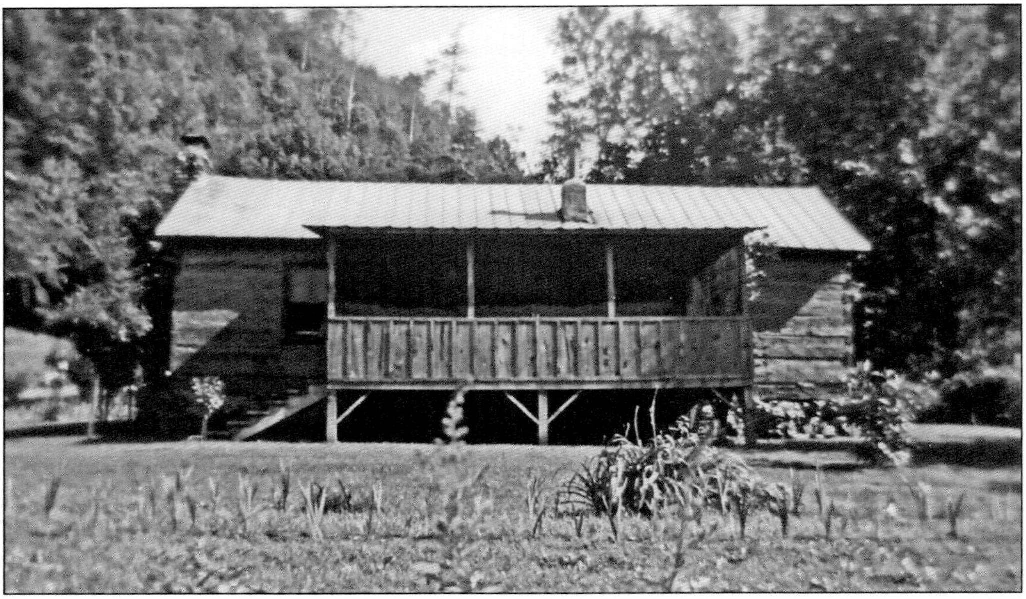

This house, owned by John W. Oliver, was rented to Ernest Ward and his family. The front porches of homes were used for many different activities; if there was a swing on the porch, many hours were spent on it. Porches were also the place for breaking beans, slicing apples, or watching the children play in the yard. (Courtesy of the William Derris Slide Collection, MS.2123. University of Tennessee Libraries, Knoxville, Special Collections.)

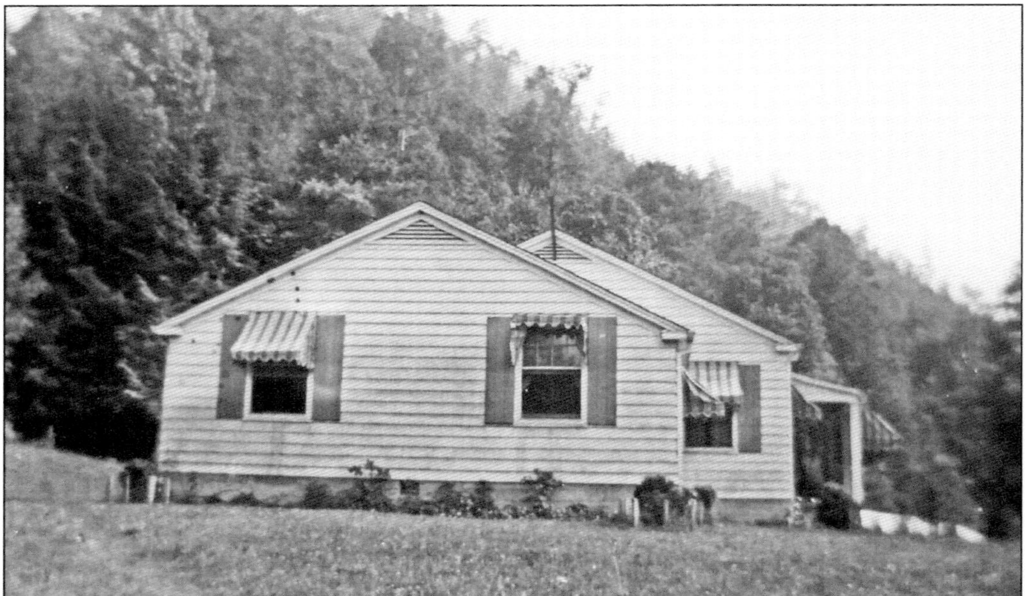

The home of John W. and Nancy Ann Whitehead Oliver stood on Bethel Church Road. John built the house in the late 1930s after moving out of Cades Cove. After moving from the cove, he continued to deliver mail there until 1939. John was a deeply religious man and passed away in 1966; Nancy Ann passed in 1948. (Courtesy of Stephen Webber, John W. Oliver collection.)

This is the barn on the John W. Oliver property on Bethel Church Road. At this time, many barns were built out of timber from the farm. In older-style barns, the upper area, called the mow or hayloft, was used to store hay and sometimes grain. A large door at the top of the ends of the barn could be opened up so hay could be put in the loft. The hay was hoisted into the barn by a system consisting of pulleys and a trolley that ran along a track attached to the top ridge of the barn. Trap doors in the floor allowed feed to be dropped into the mangers for the animals. Cows were milked daily in the barn; when it was time for milking, the cows could be seen heading toward the barn. (Courtesy of Stephen Webber, John W. Oliver collection.)

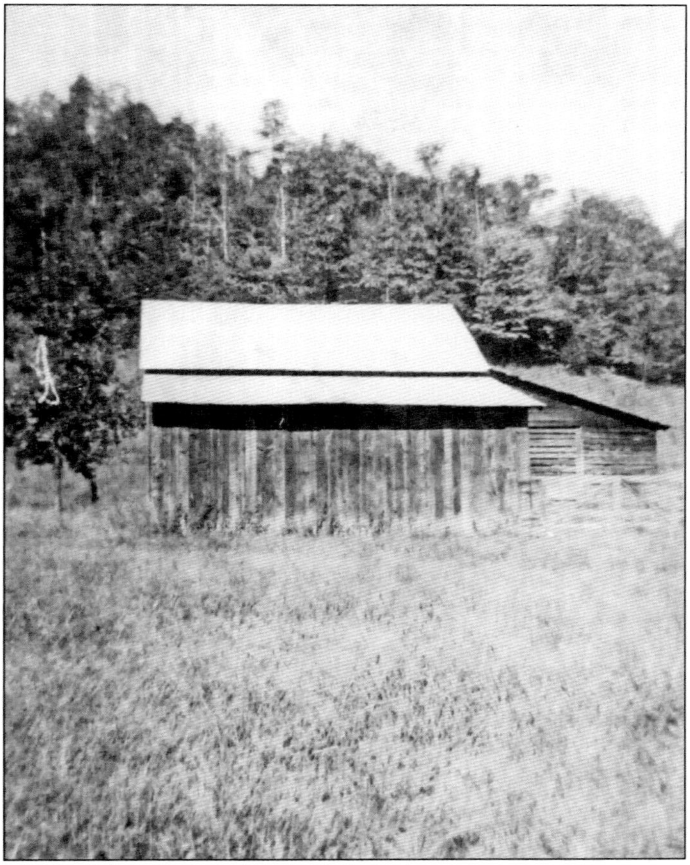

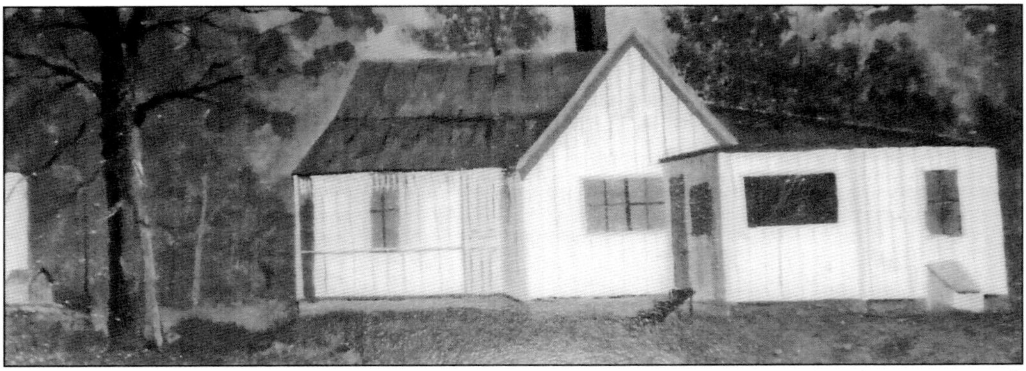

Arnold and Lou Thompson first lived in a house at the bottom of Water Tank Hill in Townsend. They eventually moved a little higher up the hill before finally moving again into this house at the top of the hill. These were "company houses," owned by D.H. "Dock" Tipton. The family living in the house at the top of the hill was responsible for taking care of the water tank for all of Tipton's houses in Townsend. Arnold worked for the Little River Lumber Mill, and he, Lou, and their children—Arnold Lee, Mary and Betty (twins), Joyce, Johnny, and Patsy—moved to the top of the hill. (Courtesy of Mary Thompson Headrick.)

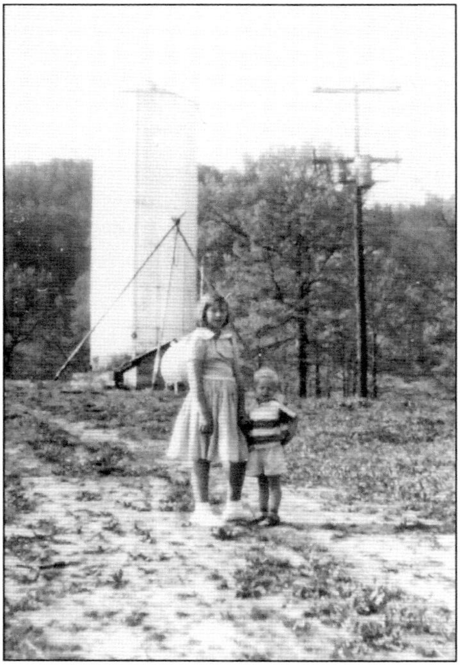

Joyce and Johnny Thompson are pictured at the Townsend water tank, which their family was responsible for taking care of. Located on their front porch was a switch with two prongs, which had to be pushed up between four other prongs. The Thompsons had to be careful to not touch the metal or they would be shocked. The water tank had an object inside called a "bug" that went up the tank as it emptied. When the bug was at the bottom, the tank was full. If the Thompsons forgot to turn off the switch, the water would run over from the spout that was located at the upper-right side of the tank. This happened many times when the children were tending to the switch and forgot. (Courtesy of Mary Thompson Headrick.)

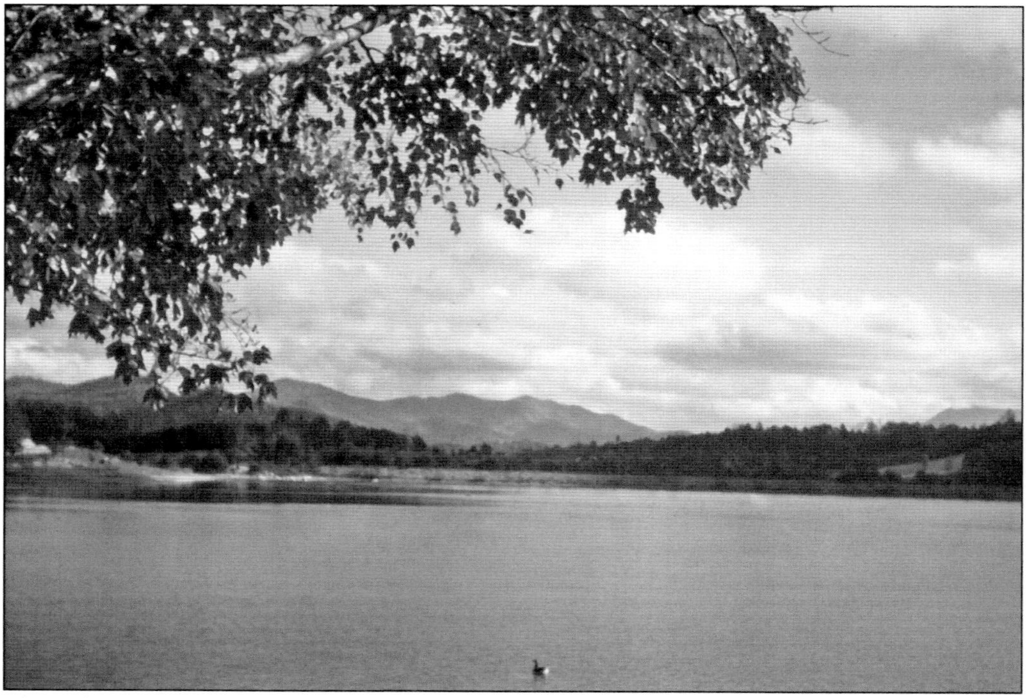

The eight-acre Laurel Lake was developed in 1937 to accommodate visitors coming to the Townsend and Kinzel Springs area. It was a wonderful place for the avid fisherman to spend the day. There was also a barrack for lodging on the property, which was used by scouts and football teams for camps. (Courtesy of the William Derris Slide Collection, MS.2123. University of Tennessee Libraries, Knoxville, Special Collections.)

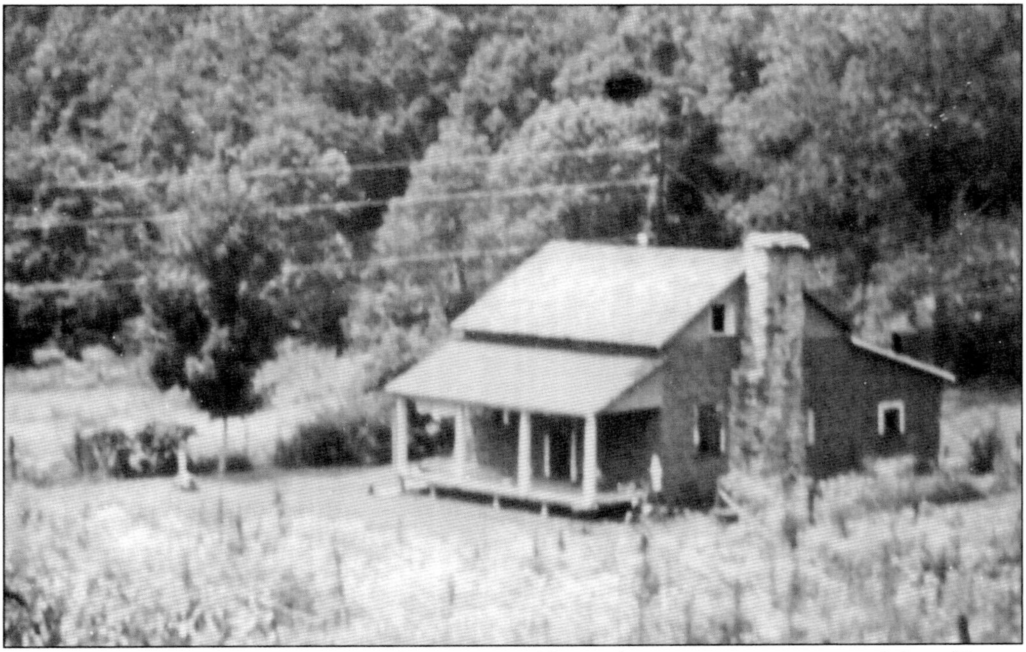

The Gibson Tipton house in Dry Valley is pictured here after some renovations and the addition of siding. The Tipton family lived here until 1962. Feeding a large family meant storing and canning vegetables grown in the garden, and the Tipton family was no different. The children also worked hard to put up foods for the winter months. (Courtesy of Louise Tipton Loan, Bertha Proctor Tipton collection.)

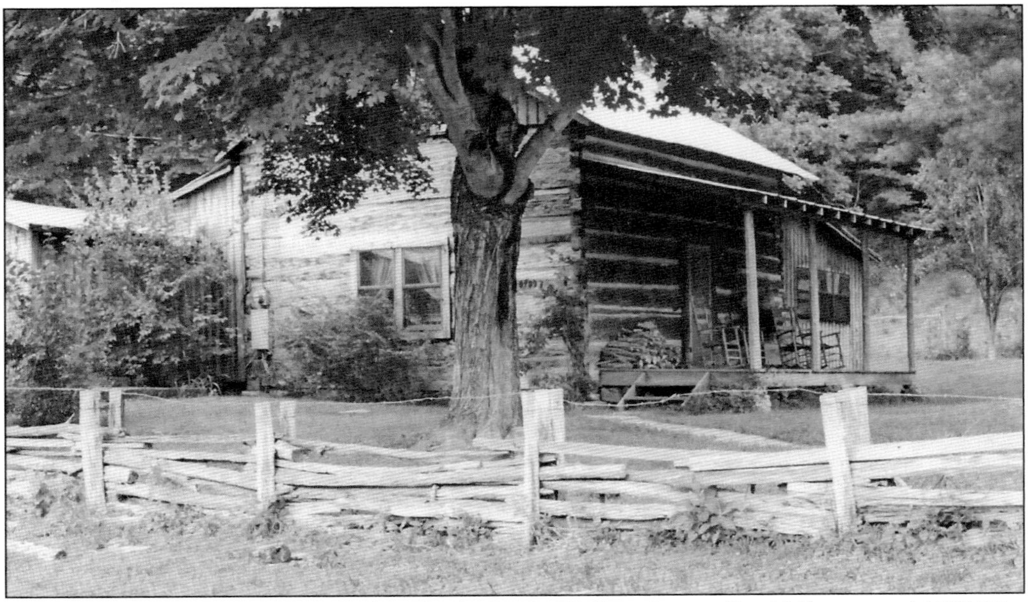

Gibson and Bertha Proctor Tipton bought the Joe L. Walker place in Dry Valley around 1935. The log cabin was probably built around the time of the Civil War. Five of the Tiptons' six children were born while they lived in Cades Cove. Daughter Louise was born while they were living in Dry Valley. (Courtesy of Louise Tipton Loan, Bertha Proctor Tipton collection.)

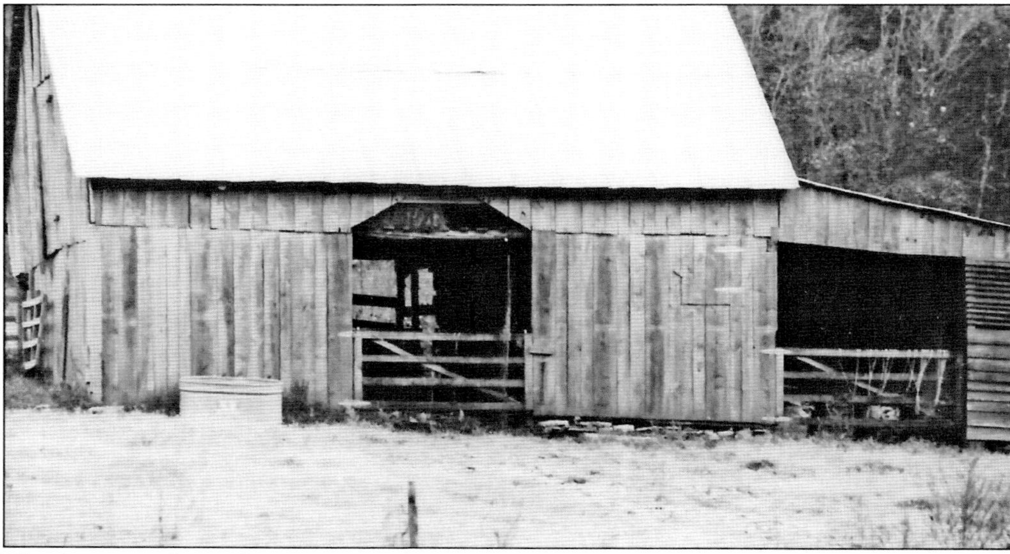

This barn stood at Gibson Tipton's farm in the Dry Valley area of Tuckaleechee Cove. Every farm would include a barn, which would keep their livestock safe from bad weather. Some had springhouses or well houses, and some would have meat-curing houses. Barns were also used as storage for the hay that was cut for the livestock. (Courtesy of Louise Tipton Loan, Bertha Proctor Tipton collection.)

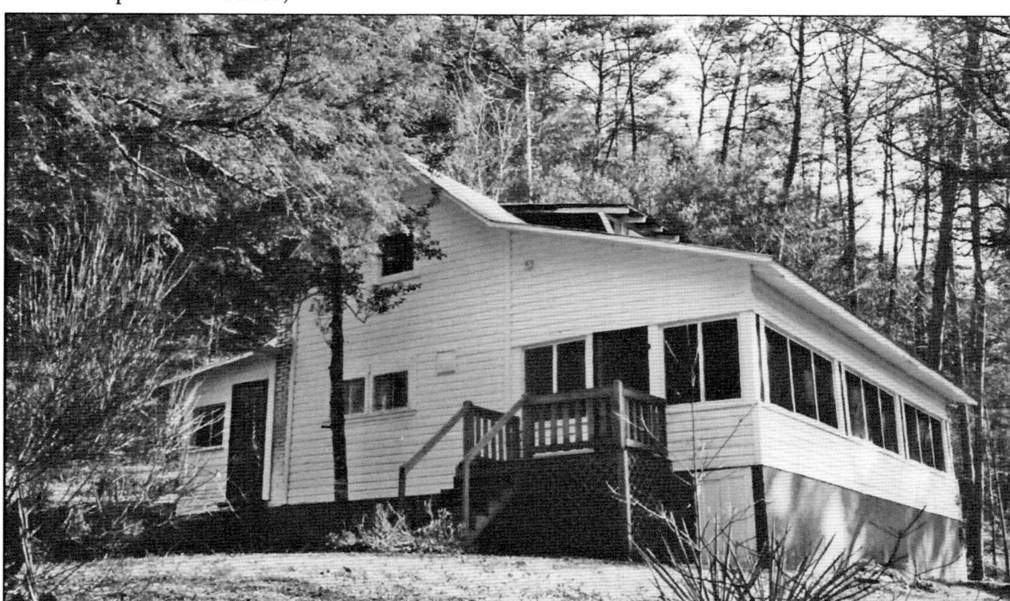

In 1962, Gibson and Bertha Tipton bought a home in the Sunshine community located in the lower part of Tuckaleechee Cove. They moved their family there from their farm in Dry Valley. Many of the homes in the Sunshine and Kinzel Springs areas were built during the era of the Kinzel Springs Hotel, when many families made their summer homes there. After the closure of the Kinzel Springs Hotel, the houses began to be sold. Gibson and Bertha lived in this house until 1992, when Gibson passed away. (Courtesy of Louise Tipton Loan, Bertha Proctor Tipton collection.)

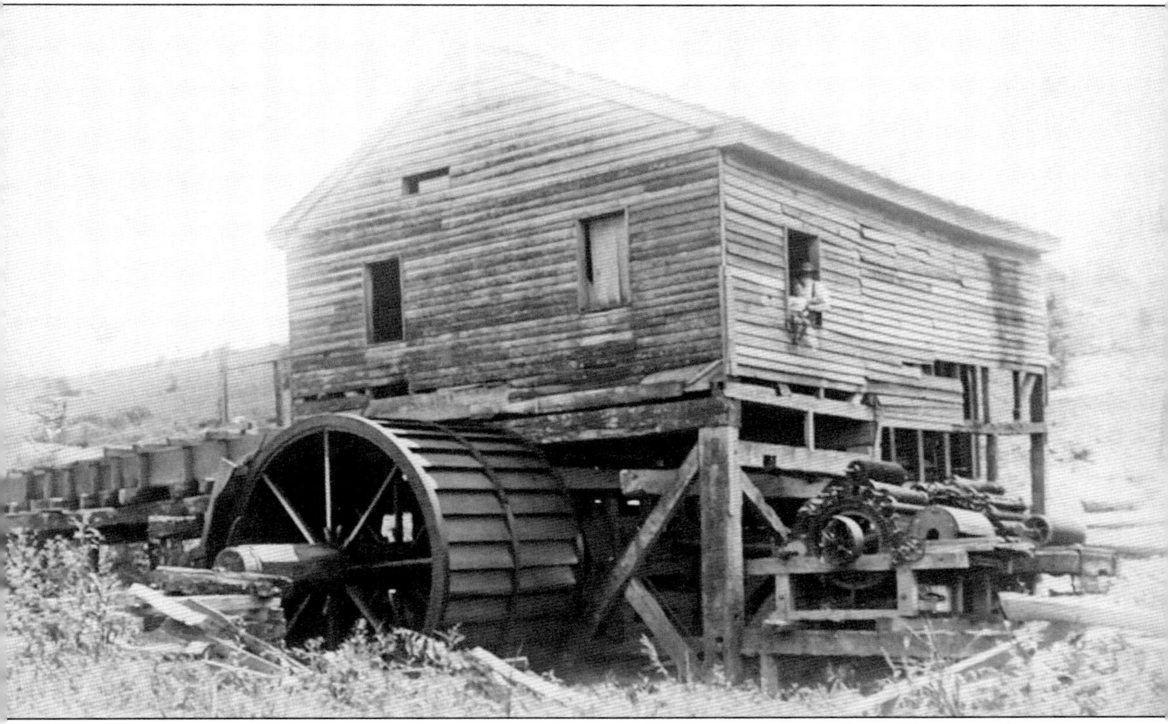

Yearout Mill on Short Creek near Dry Valley is pictured here in 1886. Early mills were almost always built and supported by farming communities. Most were operated by waterpower, with a millrace for the water to transport to the wheel. When the water hit the wheel, it turned, which then made the millstone turn. (Photograph by W.C. Cochran, courtesy of the University of Tennessee Libraries, Knoxville, Special Collections.)

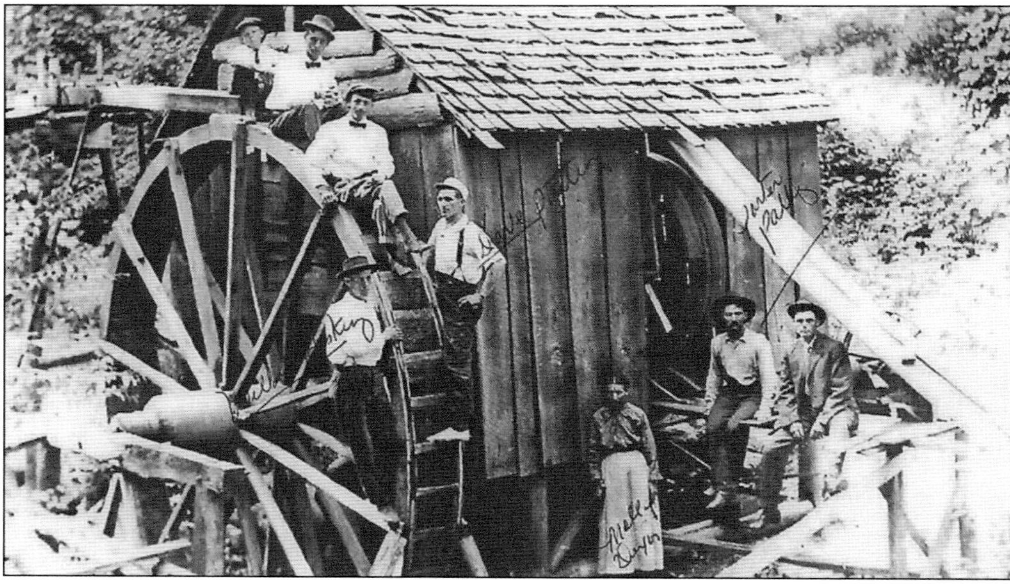

This image shows Patty Mill, located on Carr's Creek. When the grain was brought to the mill, it was lifted in the sack to be emptied into a bin, where it would then fall down through the hopper to the millstones. The milled grain was collected as it emerged through the millstone and then sent through a chute and collected into meal sacks. The charge for the service provided, called the "miller's toll," was a percentage of the grained meal. (Courtesy of Missy Tipton Green.)

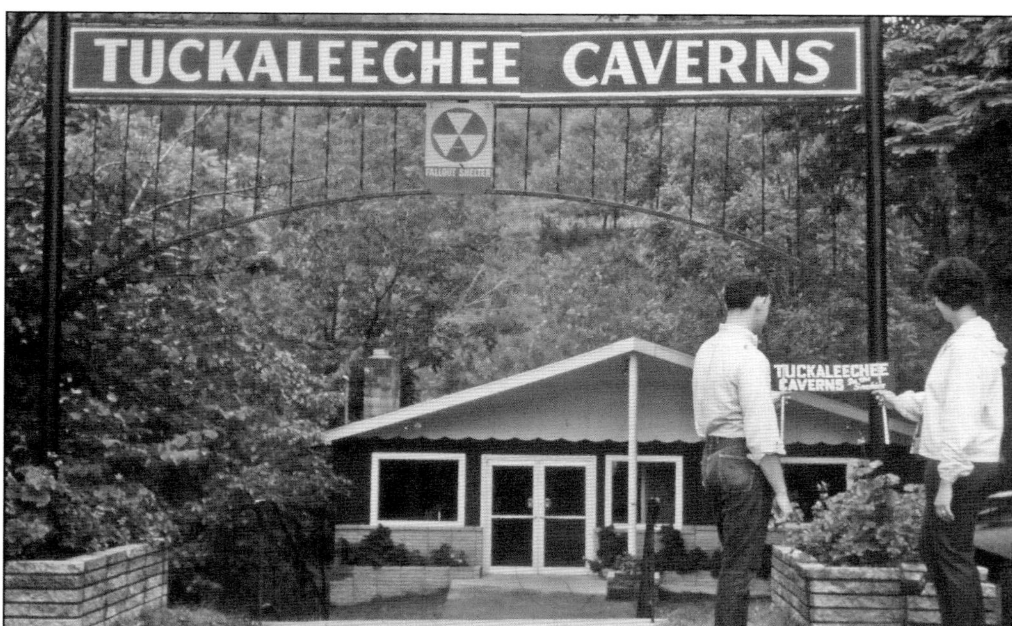

Lore has it that Tuckaleechee Caverns were first discovered in the mid-19th century when sawmill workers watched heavy rain pour into a local sinkhole. Soon, an opening was found, and the entrance to the caverns was discovered. (Courtesy of the William Derris Slide Collection, MS.2123. University of Tennessee Libraries, Knoxville, Special Collections.)

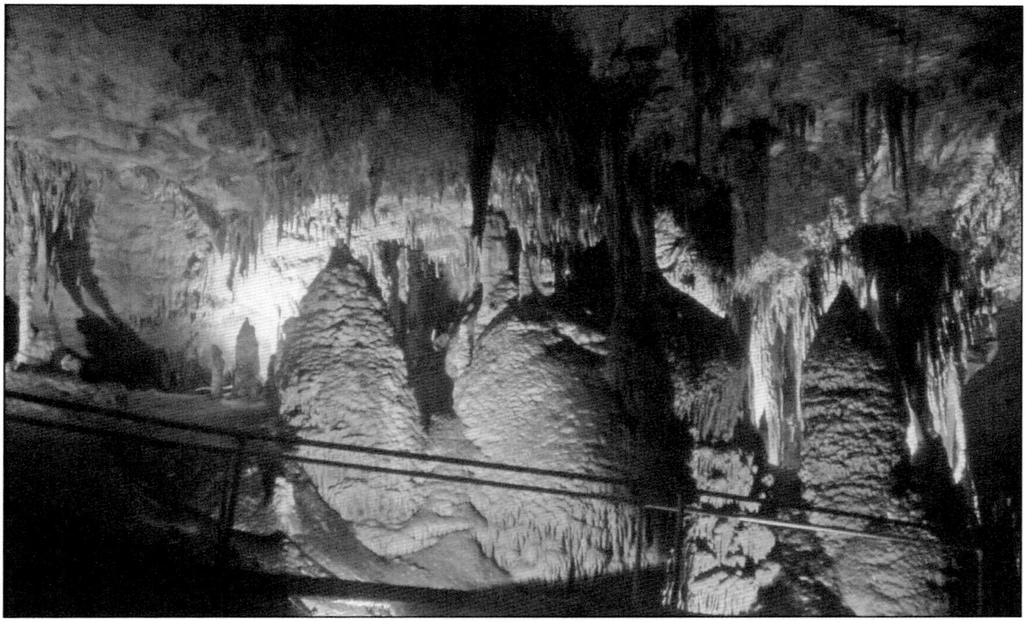

Tuckaleechee Caverns, located in the Dry Valley area, include the 200-foot Silver Falls, a double waterfall. One of the highlights of the caverns is the Big Room, which is more than 400 feet long, 300 feet across, and 150 feet deep. The Big Room was opened in 1955 for the public to view; this was also the time that electric lights were put into the caverns, replacing the long-used kerosene lanterns. (Courtesy of the William Derris Slide Collection, MS.2123. University of Tennessee Libraries, Knoxville, Special Collections.)

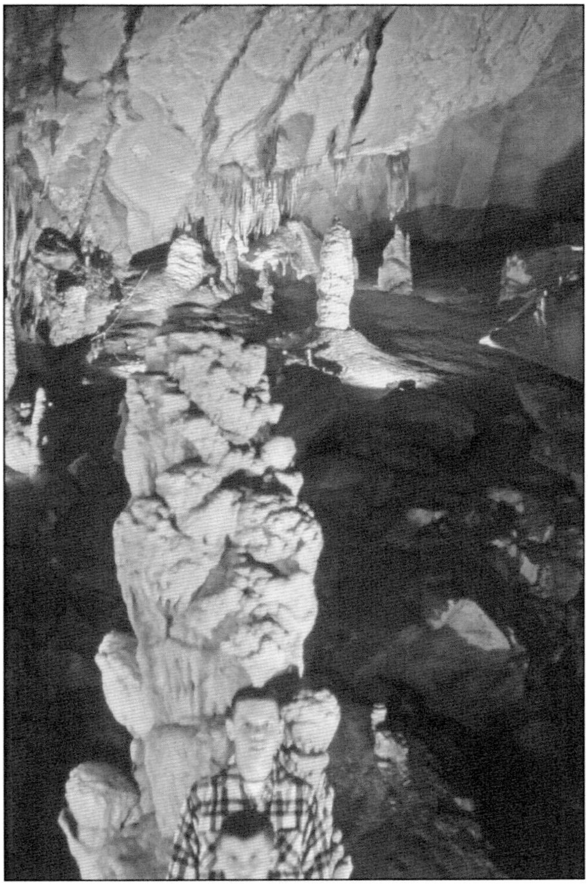

As young boys, Bill Vananda and Harry Myers often played near the opening to the caverns and ventured into them. The two men decided they wanted to open the caverns publicly and worked hard to make their dream come true. They succeeded in 1953 after carrying many tons of sand, cement, and gravel into the caverns to build steps and walkways. (Courtesy of the William Derris Slide Collection, MS.2123. University of Tennessee Libraries, Knoxville, Special Collections.)

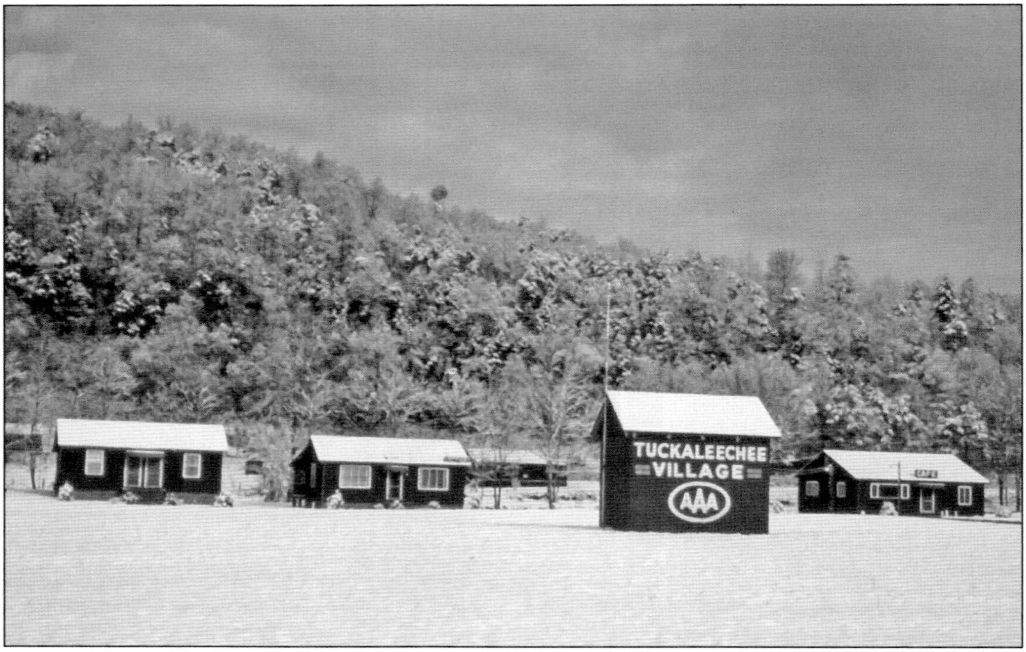

There were many vacation cabins and motels built in the Tuckaleechee Cove area. Colonel Taylor owned the Tuckaleechee Village cabins and also built the Village Barn, which was used for Saturday night music and dances. (Courtesy of the William Derris Slide Collection, MS.2123. University of Tennessee Libraries, Knoxville, Special Collections.)

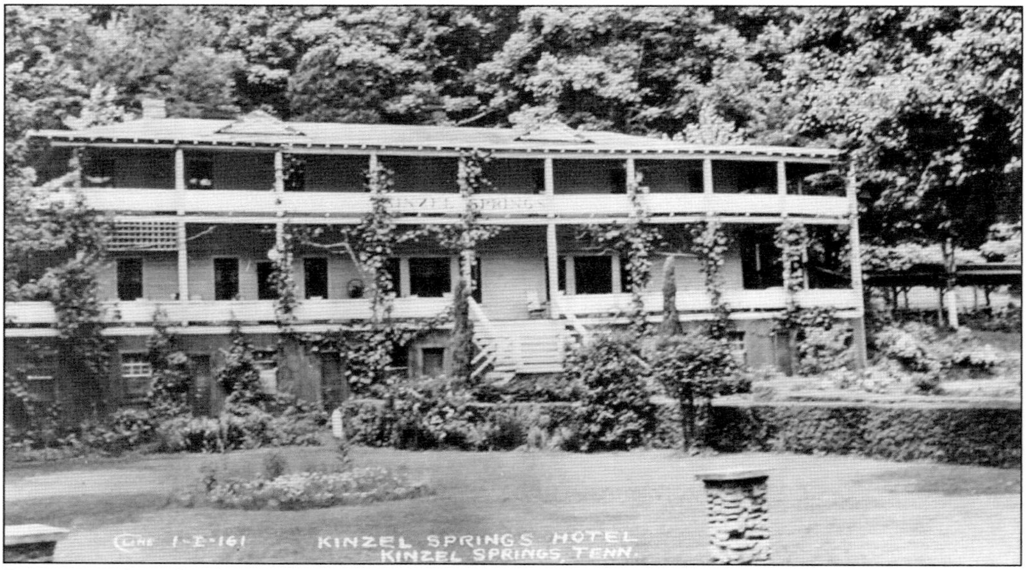

By 1914, Edward John Kinzel had built a 28-room hotel and 10 cottages bearing the name of Kinzel Springs. Access to the hotel was not a problem as Little River Railroad ran right in front of the hotel. Kinzel Springs was incorporated between 1928 and 1935. Swimming, hiking, tennis, dancing, horseback riding, and fishing in Little River were some of the activities advertised. The Kinzel Springs dining room sat 100 guests at one time and served fresh fruits, berries, vegetables out of the Kinzel garden, pure milk, and homemade butter. (Courtesy of Missy Tipton Green.)

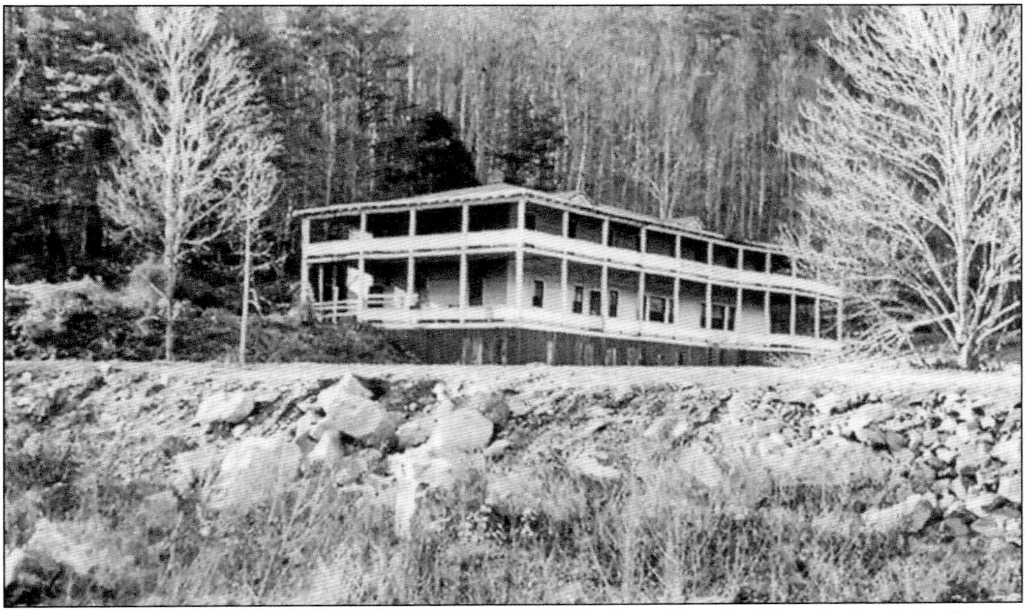

Edward John and Catherine Kinzel came into Tuckaleechee Cove looking for land. Catherine bought 29 acres from Col. J.W.H. Tipton on the west side of Little River. The tranquil beauty of the wooded area and the peaceful valley is what attracted the Kinzels to this area. The land was rich, with two mineral springs that were known for their curing waters. The Kinzels lived nearby in a cottage of five rooms that was known as Kinze-lena. (Courtesy of Missy Tipton Green.)

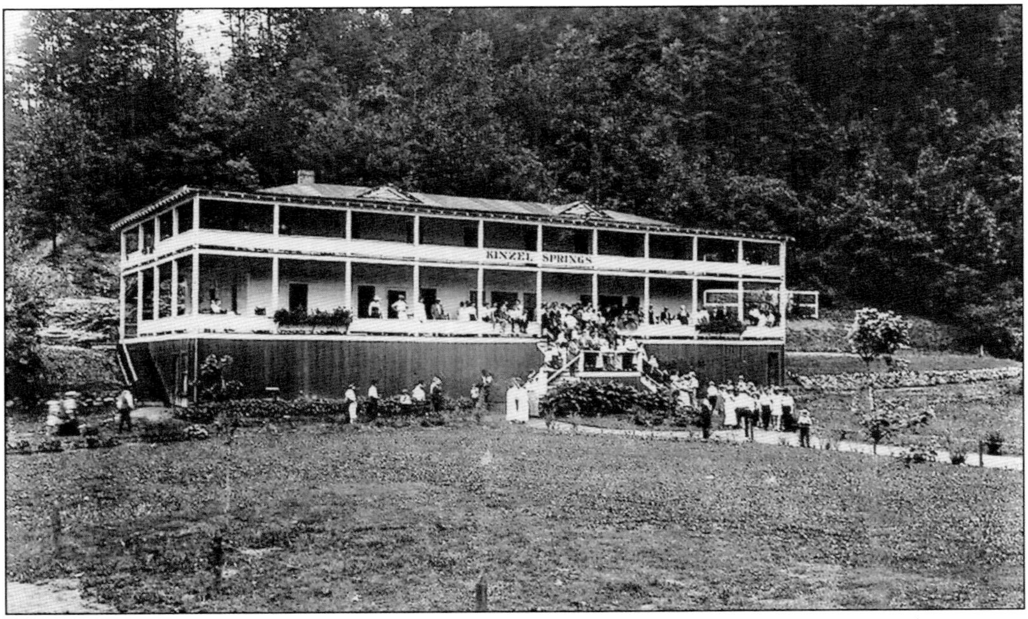

Edward John Kinzel worked hard to make the grounds of his hotel outstanding for the visitors to view. He worked tirelessly on planning his garden, which yielded fine vegetables, fruits, and flowers. The Kinzel Springs Company was formed, and many cabins were built to accommodate guests. Many families purchased land nearby and built elaborate homes for yearlong vacancy. (Courtesy of Missy Tipton Green.)

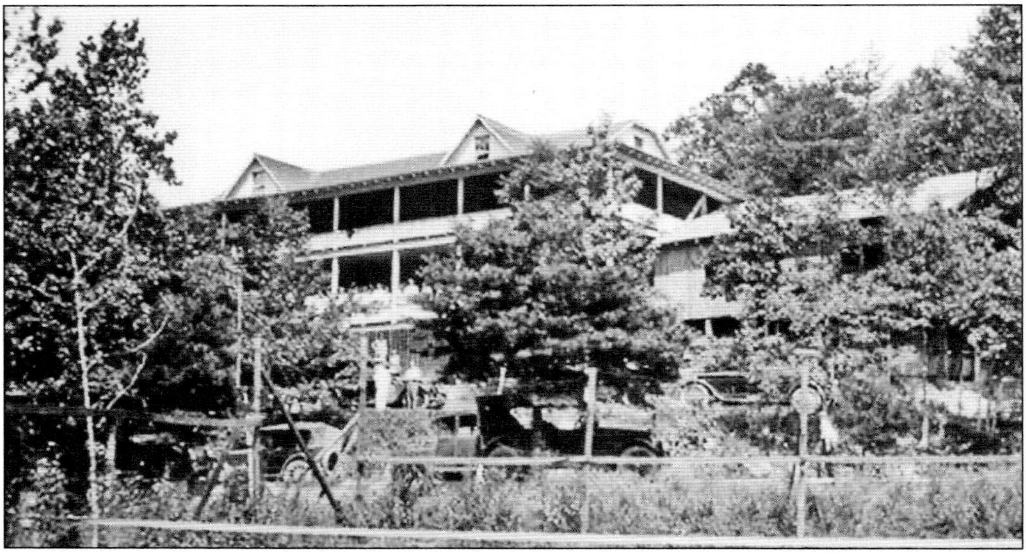

In 1903, the Kinzels bought 10 acres of land on the east side of Little River opposite Kinzel Springs. They donated a large lot for a summer home to be built for young working women from Knoxville to spend their vacations. It opened in 1907 as the Mary J. Williams branch of the Sunshine Society of Knoxville. The hotel was known as the Sunshine Rest Cottage. These young women received moderate wages and could not afford to spend vacations at an expensive summer resort. The Sunshine Rest Cottage, later known as the Smoky Mountain Inn, burned on July 21, 1952. (Courtesy of the, Albert G. [Dutch] Roth Photo Album, MS.2584. University of Tennessee Libraries, Knoxville, Special Collections.)

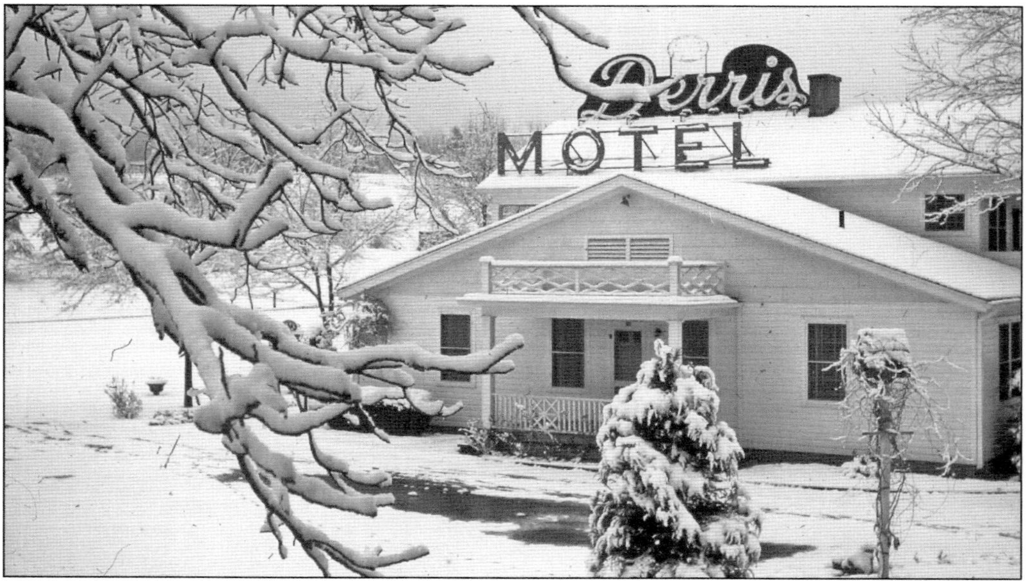

William Derris fully documented and preserved his collection, which is now stored at the University of Tennessee Library Special Collections. He visited many Townsend residents and sometimes guided his guests on nightly tours of the area and took many photographs. (Courtesy of the William Derris Slide Collection, MS.2123. University of Tennessee Libraries, Knoxville, Special Collections.)

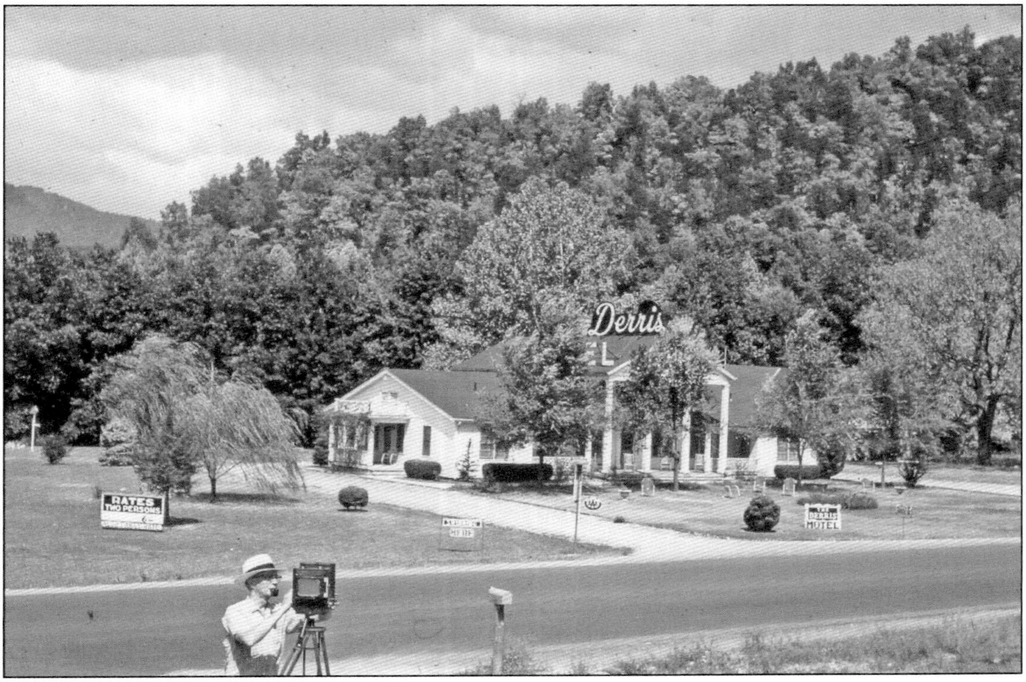

William Derris was the owner of the Derris Motel in Townsend, which was purchased from the Bashor family in the early 1940s. Derris photographed guests, seasonal landscapes, flora, wildlife, pets, and the surrounding areas of East Tennessee. An avid amateur photographer between the mid-1940s and late 1960s, he showed guests slide presentations of his wonderful pictures. These slides were done in color, which was uncommon for this time. (Courtesy of the William Derris Slide Collection, MS.2123. University of Tennessee Libraries, Knoxville, Special Collections.)

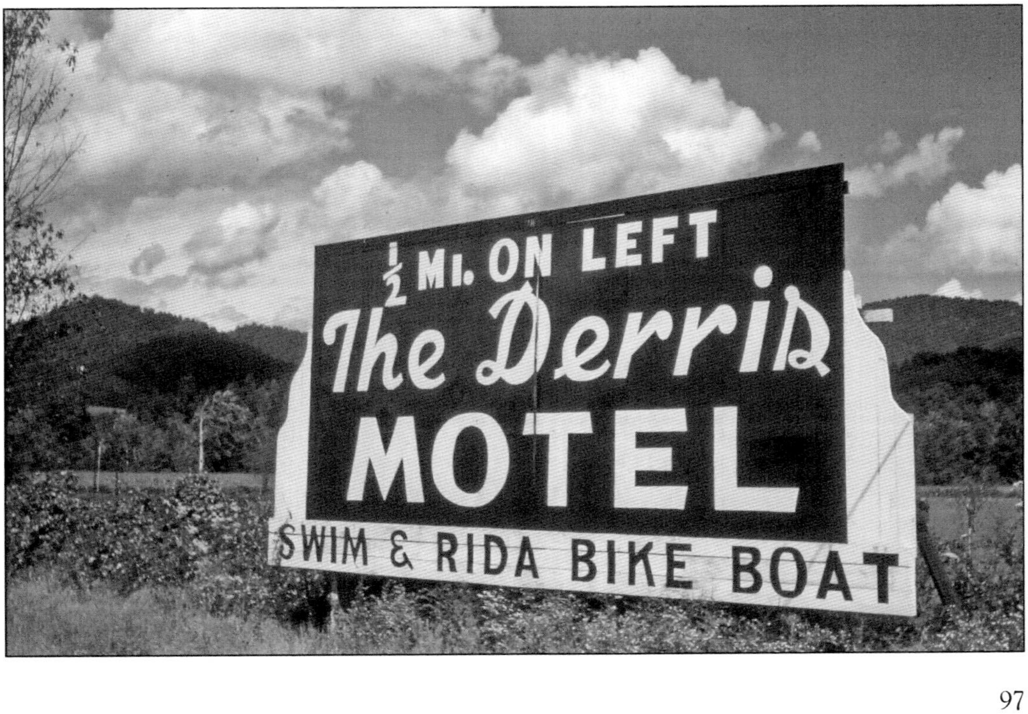

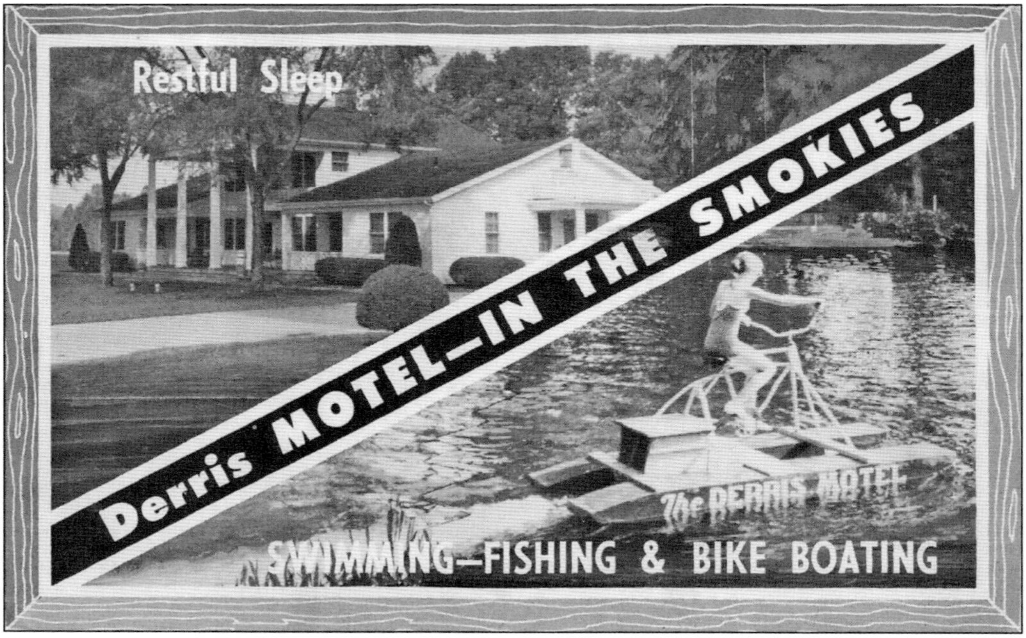

Derris Motel was known for its bike boating rides in the river behind the motel. Many of the young people of the Townsend area worked for Derris in the summer when school was out. (Courtesy of the William Derris Slide Collection, MS.2123. University of Tennessee Libraries, Knoxville, Special Collections.)

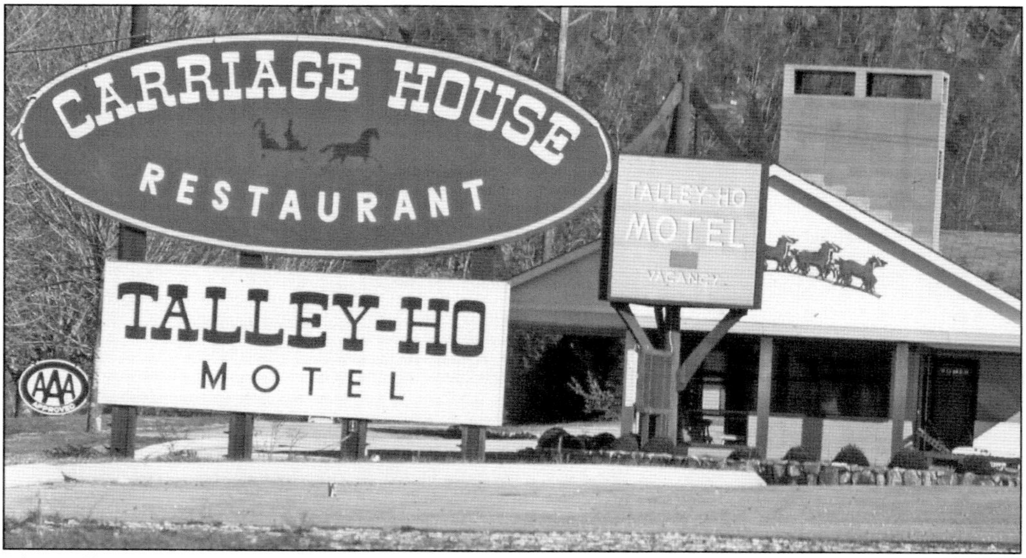

The Talley-Ho Motel was built in 1953 by Victor and Helen Talley. Victor worked very hard to provide guests with the best customer service he could. He wanted to make sure their trip to Townsend was an experience they would not forget. The Talley-Ho is still operated by the Talley family today. After the closure of the Little River Railroad & Lumber Company, Townsend became the entrance to the Great Smoky Mountains National Park. A number of motels were built in this area to accommodate the many visitors through this area. (Courtesy of the William Derris Slide Collection, MS.2123. University of Tennessee Libraries, Knoxville, Special Collections.)

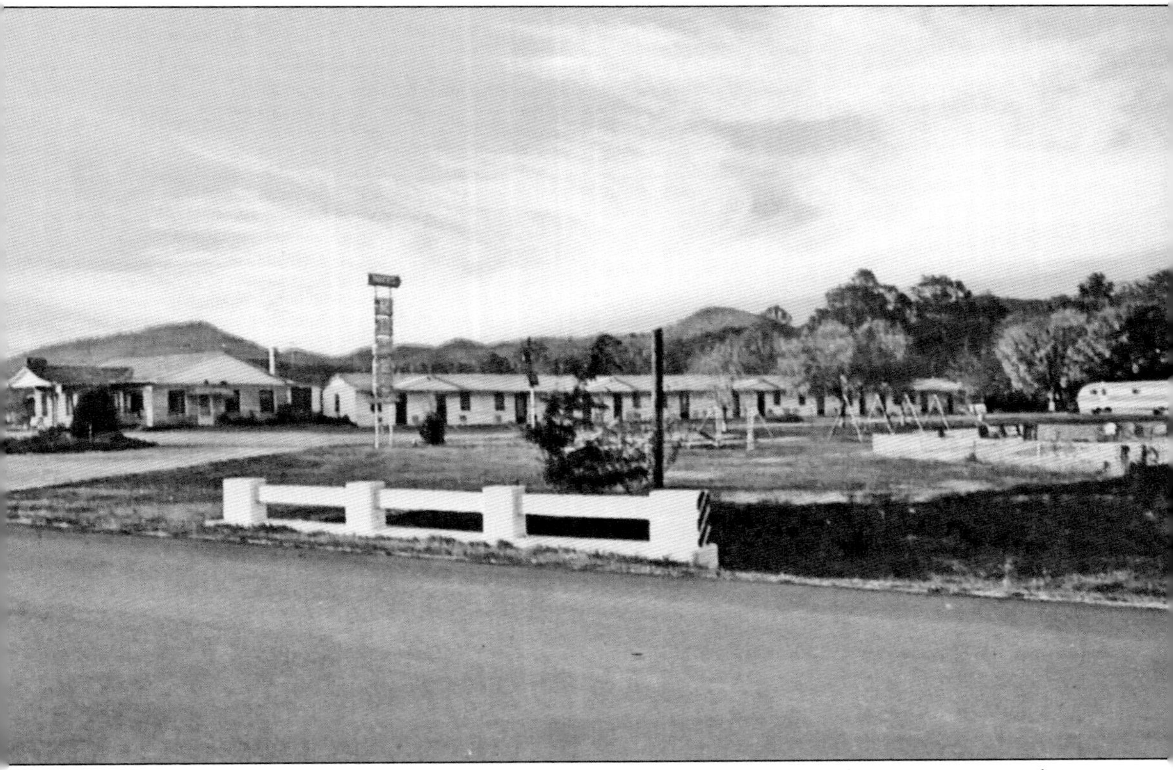

Dock's Motel was another early motel in the Townsend area that was built in response to the extensive travel of tourists to Great Smoky Mountains National Park. It was constructed in 1935 by Lorene Whaley Huskey. Huskey was from the Greenbrier area, which is now part of Great Smoky Mountains National Park. The original location of Dock's cabins sat close to the entrance to the park. Dock's had to relocate in the 1950s when the state obtained the right-of way for the current road. The site now occupied by Dock's was the Sam Law place where an old, two-story, white house used to sit. The building began as cabins that were later converted into the motel. Little River Railroad passed by in front of the motel and dropped off logs that were cut and planed right on site then taken into the cabin and nailed up, all in one day. (Courtesy of Missy Tipton Green.)

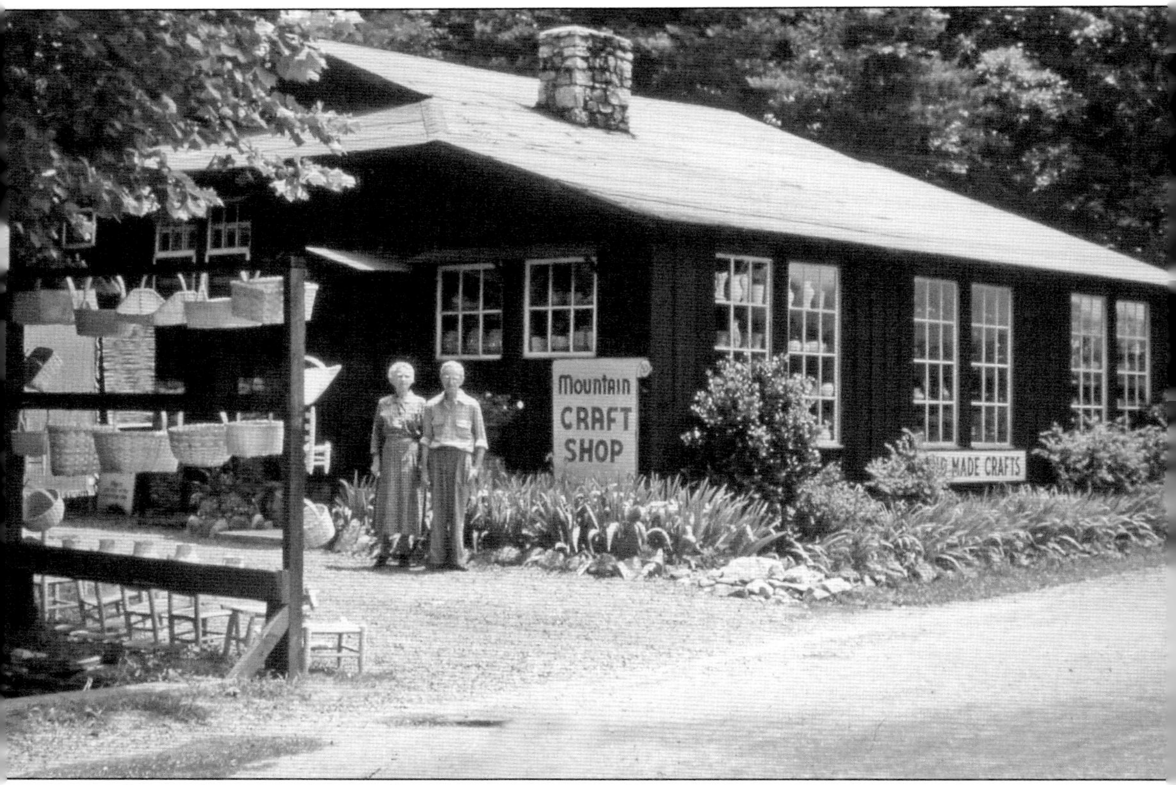

This craft store in the Sunshine area of Townsend, pictured in 1953, was operated by Mr. and Mrs. Frank Jones. This store offered crafts made by the local people of Tuckaleechee Cove. (Courtesy of the William Derris Slide Collection, MS.2123. University of Tennessee Libraries, Knoxville, Special Collections.)

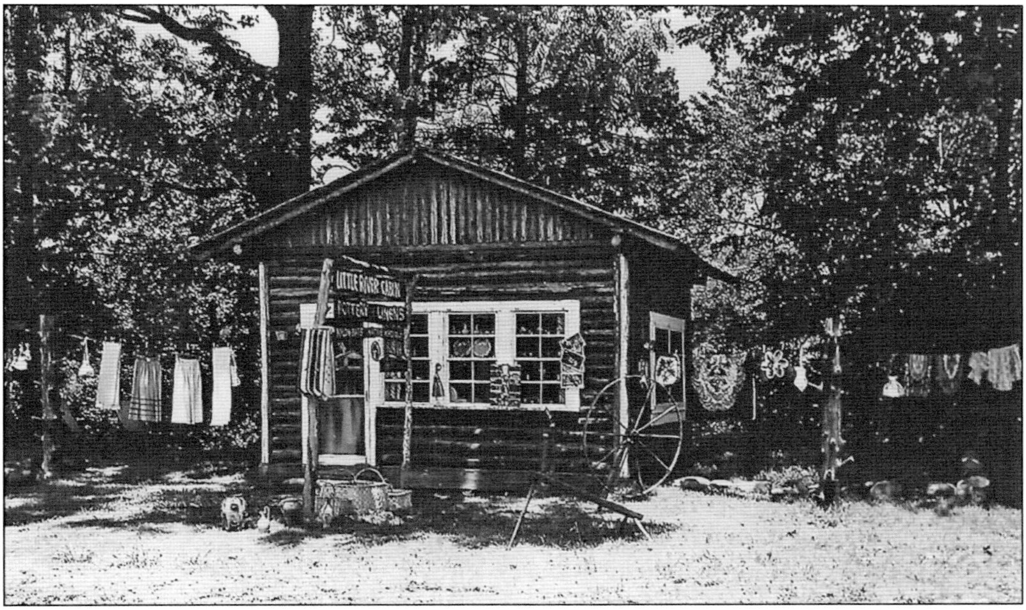

The sign in front of the Little River Cabin in the Kinzel Springs area of Tuckaleechee Cove says pottery and linens are for sale. Items made by residents of this cove could be sold here to the tourists coming into Kinzel Springs Hotel. Some of the wares sold are shown hanging on the lines, and a vintage spinning wheel sits in front of the building. (Courtesy of Marilyn Myers Byrd Photograph Collection.)

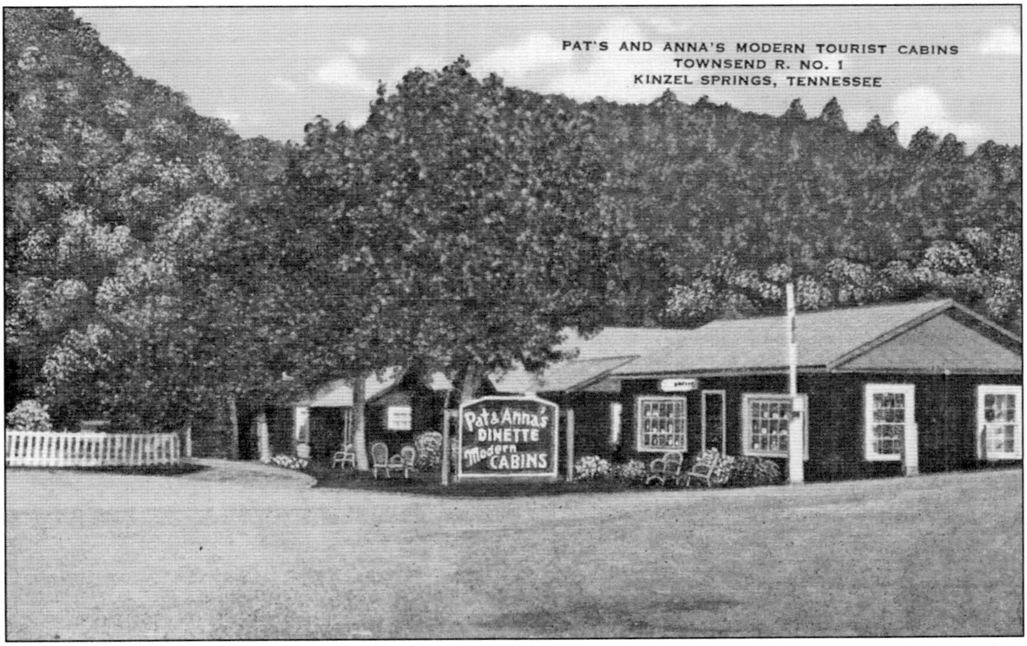

John and Norma Hitch Wilson ran the Pat & Anna's restaurant for about five years. This restaurant was located in the Kinzel Springs area of Townsend and was known for good country food. As the picture shows, there were tourist cabins as well as a restaurant. (Courtesy of Missy Tipton Green.)

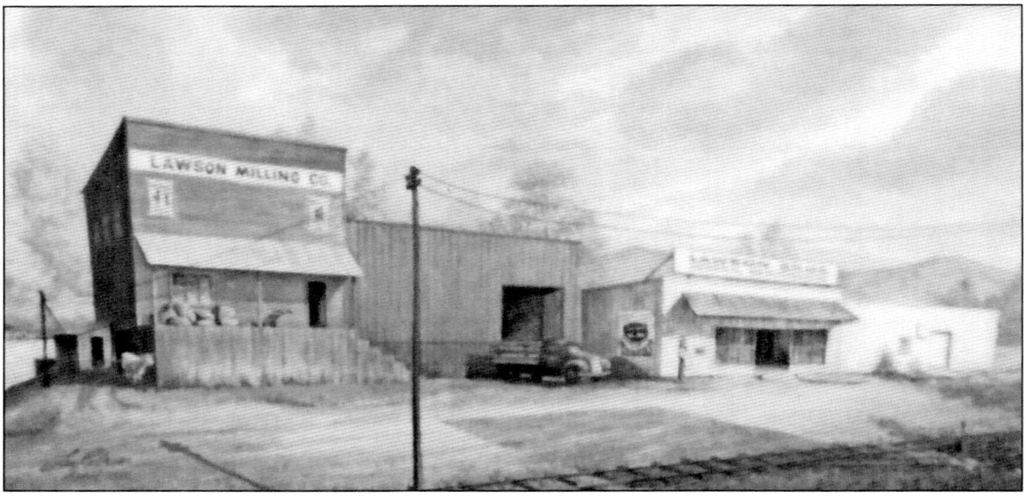

Lawson Milling Co. and the Lawson Bros. store were located close to the intersection of Webb Road and East Lamar Alexander Parkway. William Tobias Lawson built a grain mill on the banks of Little River. A few years later, a general store was built next door. William's son Arville Raulston Lawson had many stores—one in Wears Valley, one in Grainger County, and one in Rockford. In 1938, Arville and his wife, Ila, bought into the Townsend store with Arville's parents, and Ila managed it full time. The mill was operated until 1950. People would bring their corn in horse-drawn wagons from valleys and coves surrounding Townsend; they would be lined up 10 or 15 deep. When the store closed in the 1960s, the Lawsons wrote off thousands of dollars of debt. (Courtesy of Fred Lawson.)

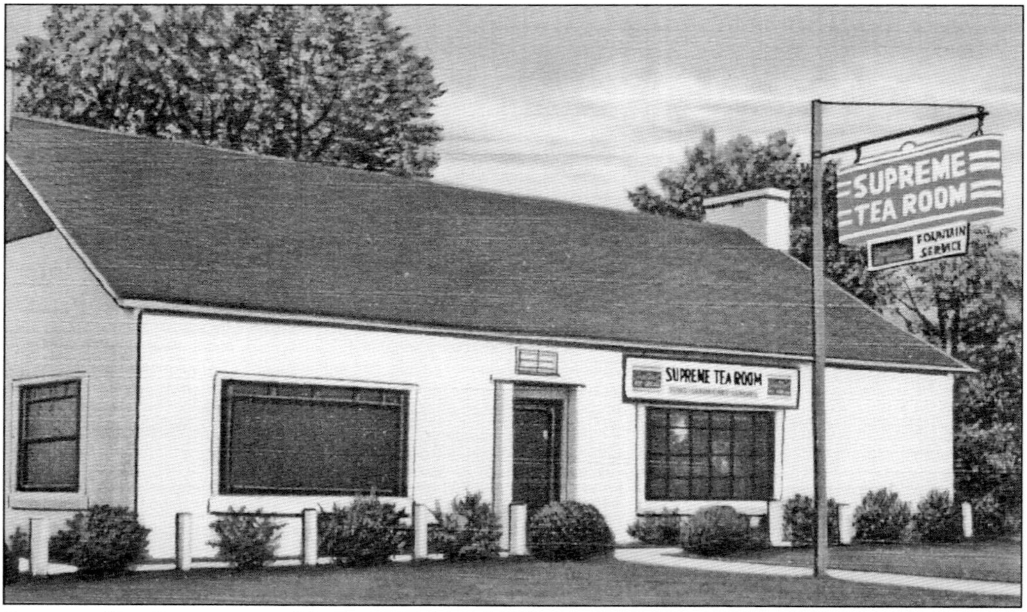

Supreme Tea Room, owned by Joe Hauk and Lillian Myers, was operated by Col. W.C. Taylor around 1947. The Supreme Company in Knoxville, which Taylor operated, manufactured the Supreme lemony mayonnaise and salad dressing. The building is now the location of the Townsend Shopping Center. (Courtesy of Missy Tipton Green.)

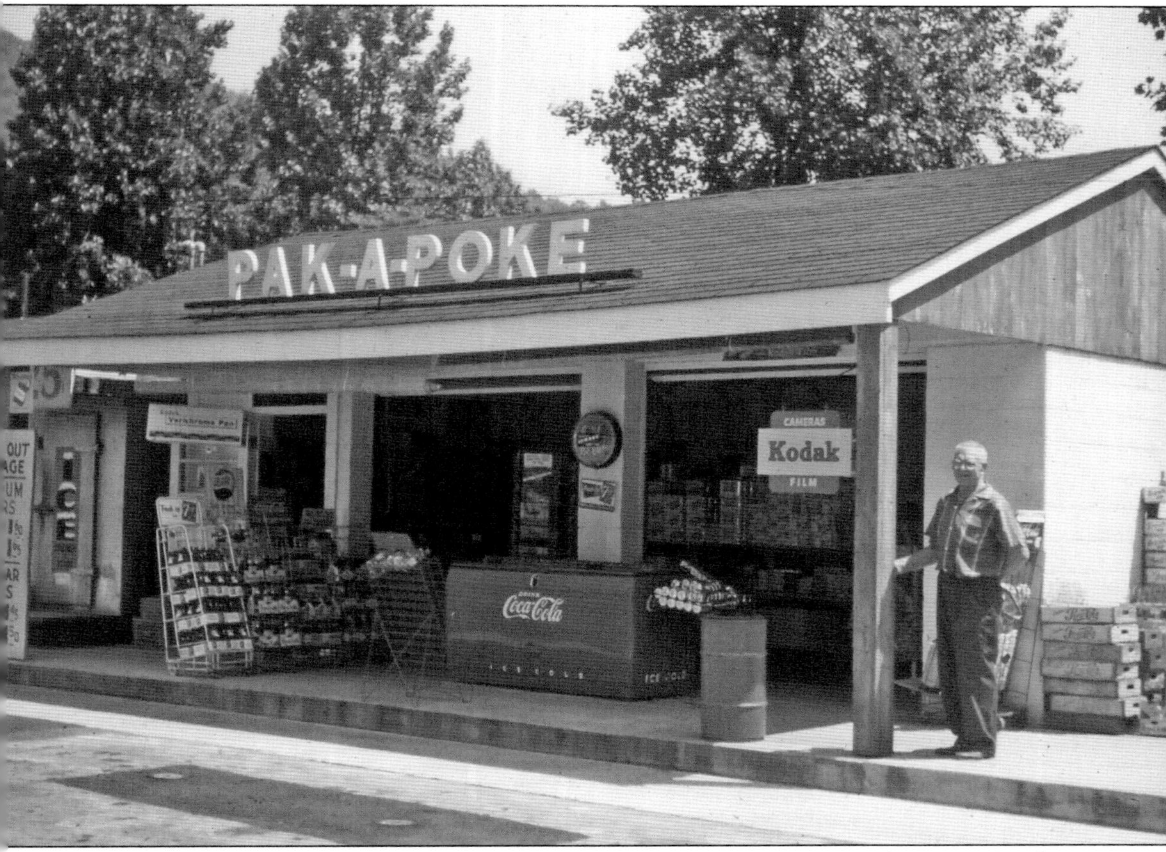

Reginald Blair's Pak-A-Poke store was the last stop before the entrance into the Great Smoky Mountains National Park. Many of the young people living in the Townsend area would get jobs there. (Courtesy of the William Derris Slide Collection, MS.2123. University of Tennessee Libraries, Knoxville, Special Collections.)

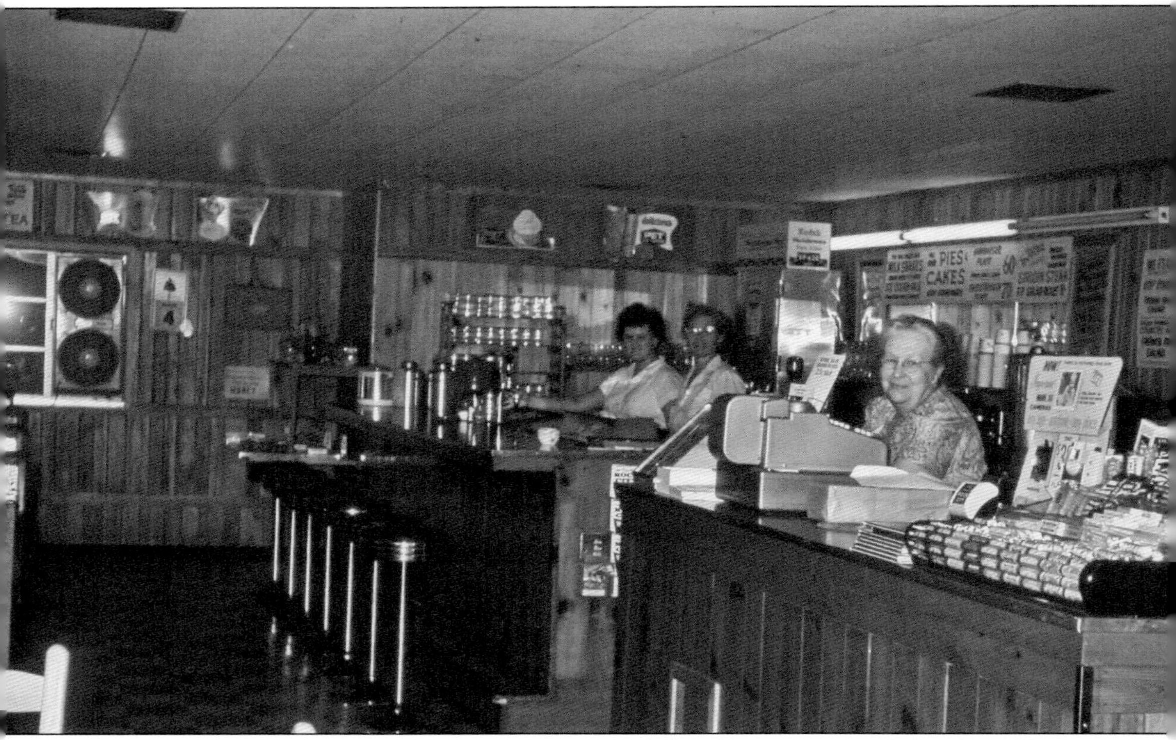

Blair's Restaurant, pictured on July 30, 1958, was a full-scale restaurant with booths, tables, and a soda fountain for making milk shakes, sodas, and Coke floats. There was also a gift shop inside. The restaurant sat next door to the Pak-A-Poke store. (Courtesy of the William Derris Slide Collection, MS.2123. University of Tennessee Libraries, Knoxville, Special Collections.)

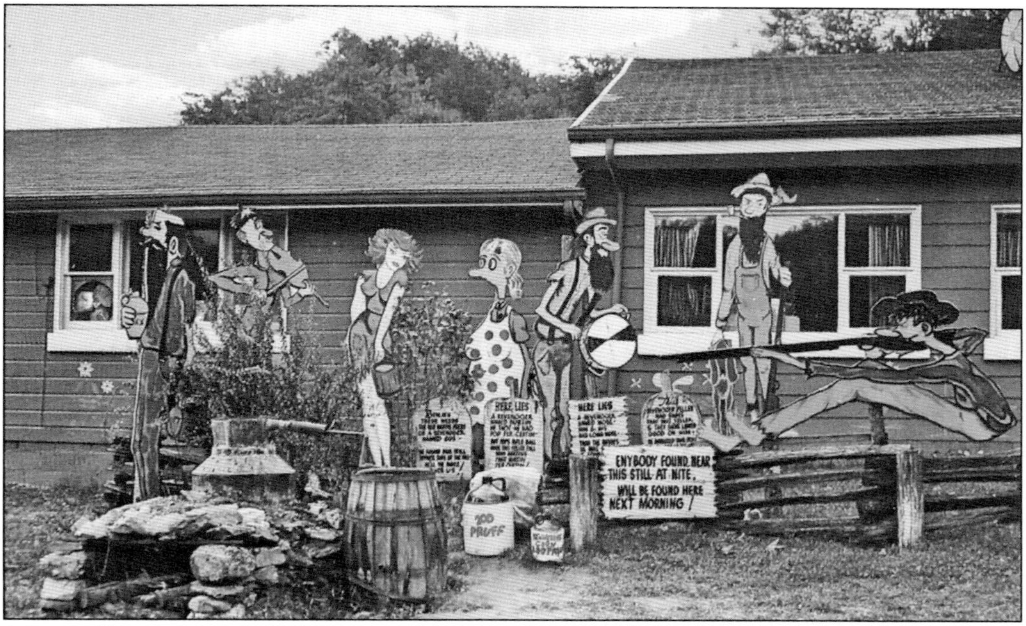

John and Norma Hitch Wilson operated the Wilson's Hillbilly Restaurant on Old Highway 73 (now known as Highway 321) for 15 years. It was known for serving some of the best country cooking and homemade bread. Norma promoted a successful advertising program for the restaurant depicting herself and her husband as comic hillbillies. The Wilson's Hillbilly Restaurant was open from 1945 to 1958. (Courtesy of Missy Tipton Green Photograph Collection.)

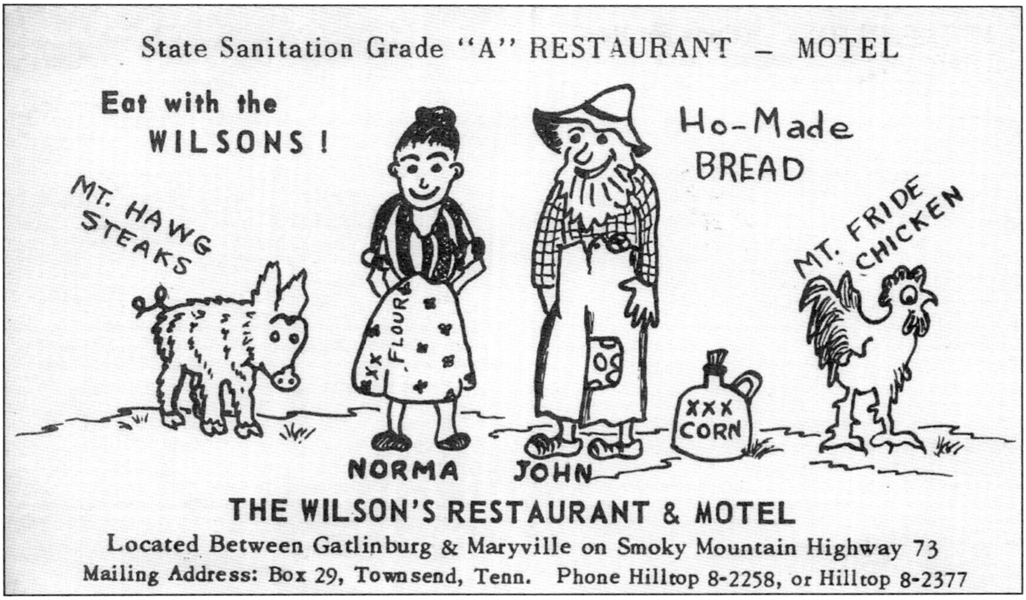

Norma Wilson was also quite an authority on mountain lore and did much to help residents of other areas appreciate the fine qualities of the people who lived in Tuckaleechee Cove. She authored many booklets dealing with mountain recipes, cures, and folklore. She was responsible for promoting tourist travel to Townsend and Tuckaleechee Cove. (Courtesy of Missy Tipton Green Photograph Collection.)

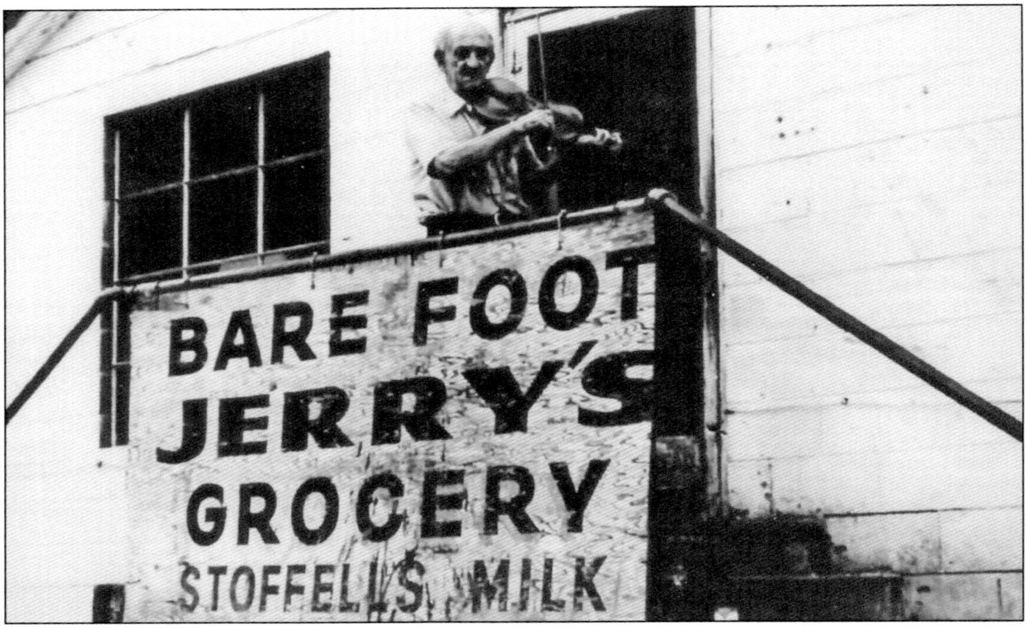

Barefoot Jerry Effler learned to play the fiddle as a young boy; he played by ear without reading music. In 1970, a group of bluegrass and rock musicians, who went by the name of the Charlie Daniels Band, rented a cabin behind Jerry's store. They spent much time with Jerry and came to know him quite well. They released an album with Jerry's store on the front, and as the album jacket explains, "They had come to feed off of the mountain magic and feel the juices of their creativity stir and emerge through a natural environment." (Courtesy of Paulette Ledbetter.)

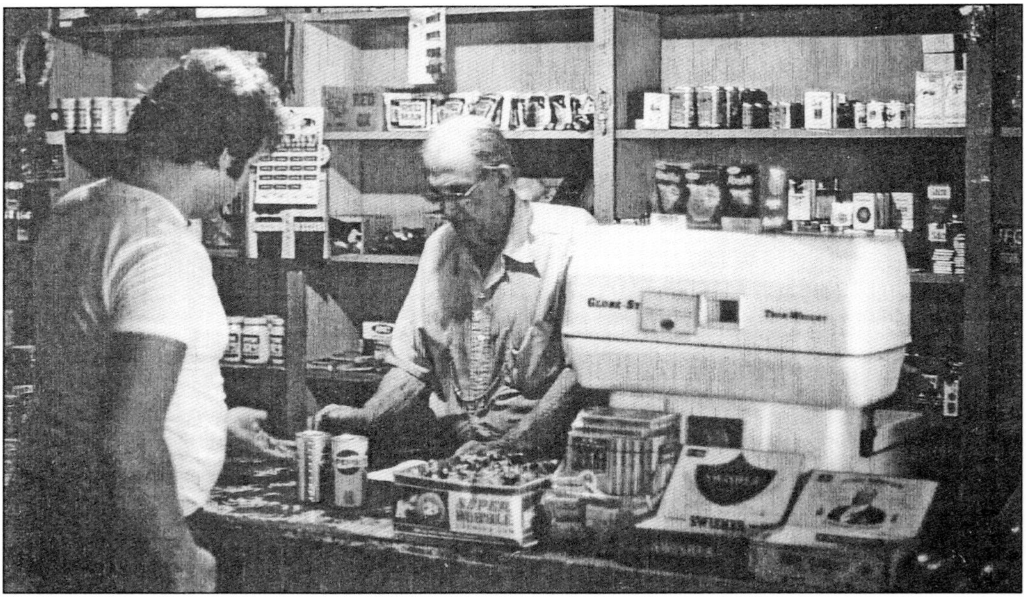

Here, Effler is helping a customer inside his country store. Born Jerry Osborne Effler in 1880 in El Dorado, a settlement between Cades Cove and Townsend, he came by the name "Barefoot" because he would play his fiddle and dance for his customers without any shoes. (Courtesy of Paulette Ledbetter.)

Four

Little River Railroad and Little River Lumber Company

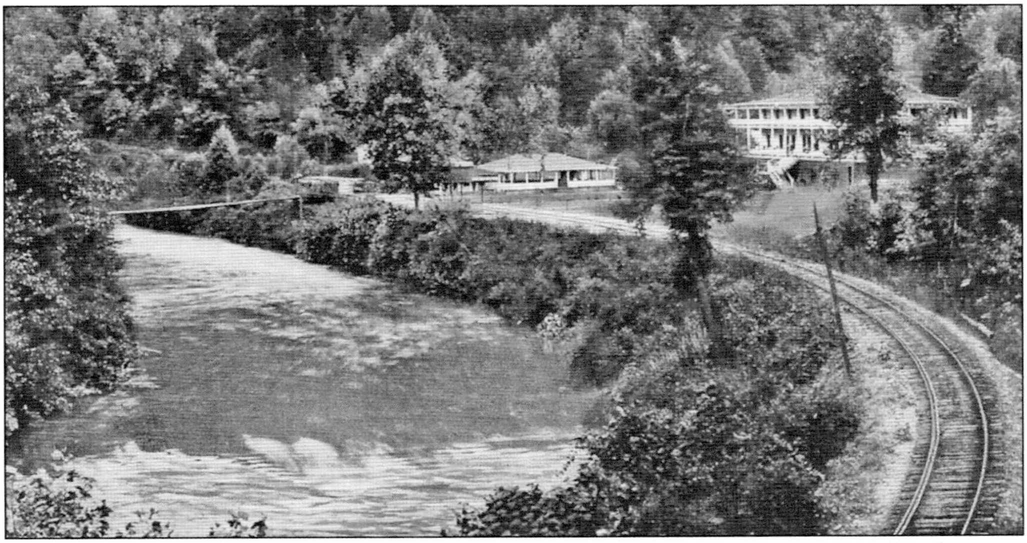

Here, Little River and the railroad pass by Kinzel Springs Hotel. The Sunshine railroad station is in front of the hotel, and the swinging bridge leading over to the Sunshine area is in front of the station. On the Sunshine side of the river was the Old State Route 73, which led into the Townsend area. (Asheville Postcard Company, Asheville, NC, No. 69931.)

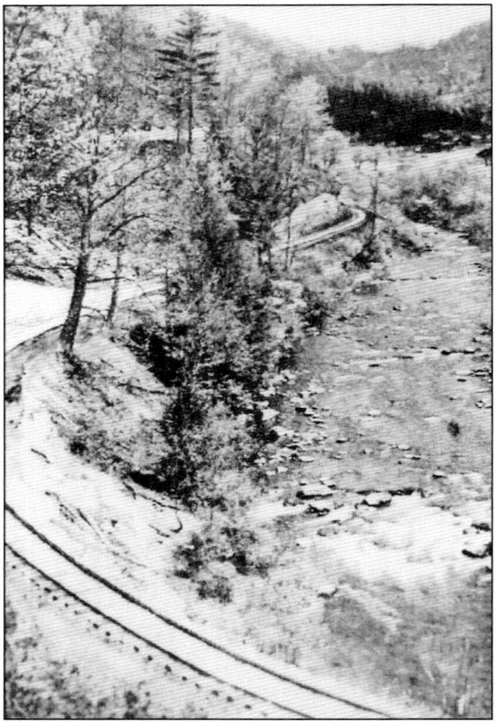

The Schlosser tannery in nearby Walland needed bark for the leather-tanning business. The Walton & England Company–Schlosser Leather Company already had the railroad constructed from Maryville to Walland. A railroad was needed to ship the raw hides to the tannery and then ship out the finished product. (Courtesy of Missy Tipton Green Photograph Collection.)

In this 1920 photograph are D.H. "Doc" Tipton, Stuart McNiell Sr., Tom Carver, Clyde LeQuire, Flo Dew, and W.B. Townsend. Tipton was the superintendent of Little River Lumber Company; McNiell, the bookkeeper; Dew, the secretary; and Townsend, the founder. (Courtesy of McNiell Family Collection.)

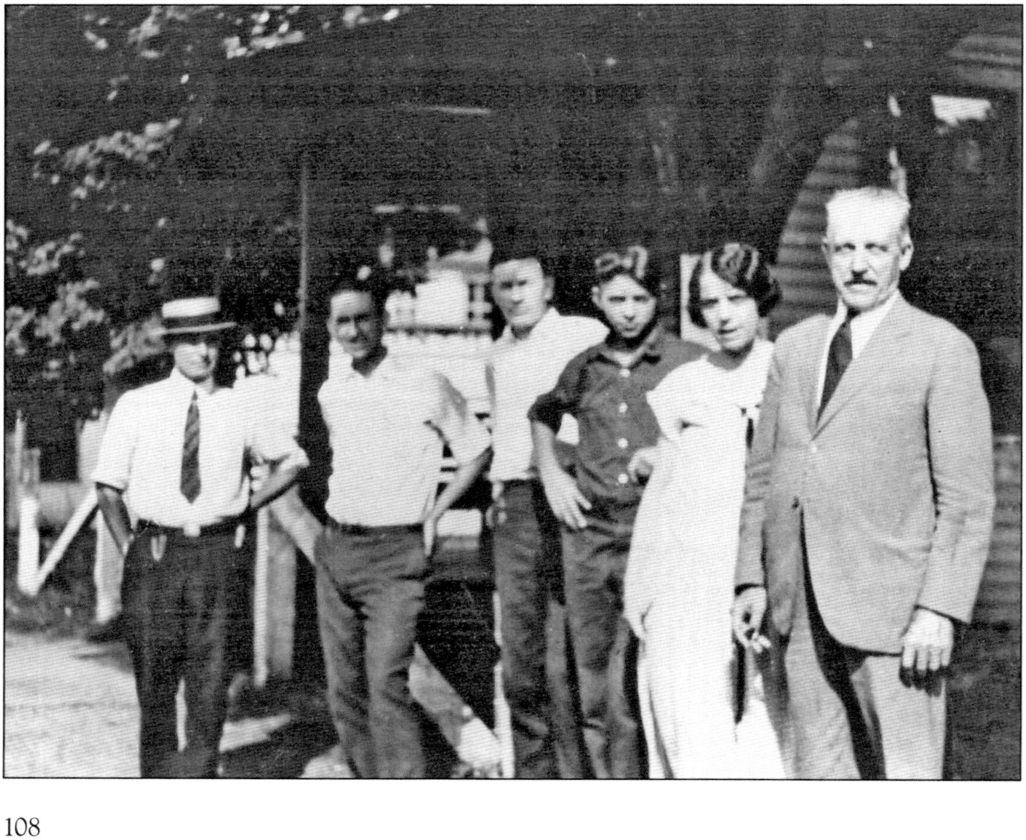

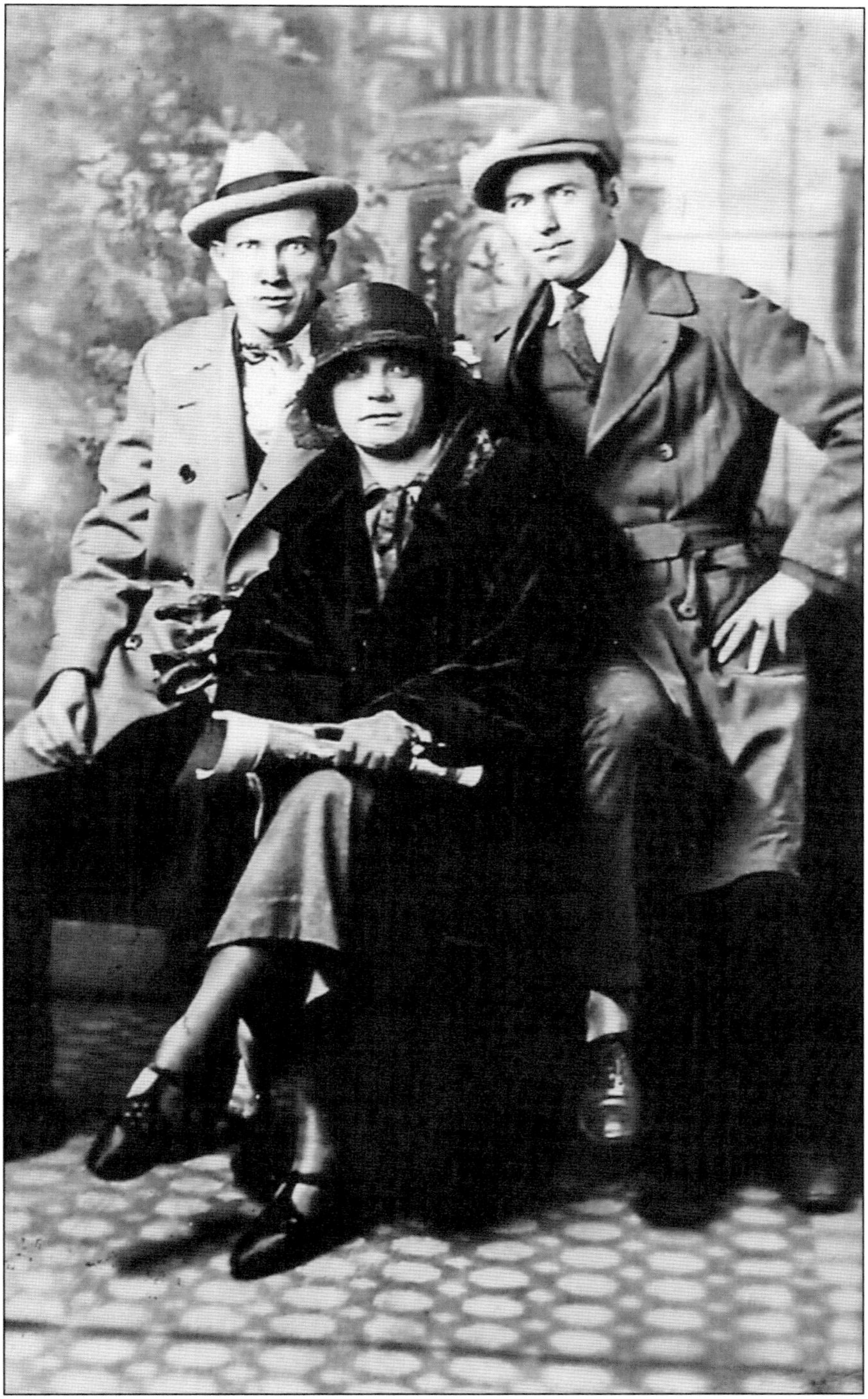

Woody and Flo Dew are pictured with Stuart McNiell Sr. (right) in 1924. Woody was an engineer for the Little River Railroad, while his wife, Flo, worked as a secretary. Stuart McNiell Sr. was the bookkeeper. In the early years of the lumber company's operation, the company marketed 57 varieties of hardwood. (Courtesy of McNiell Family Collection.)

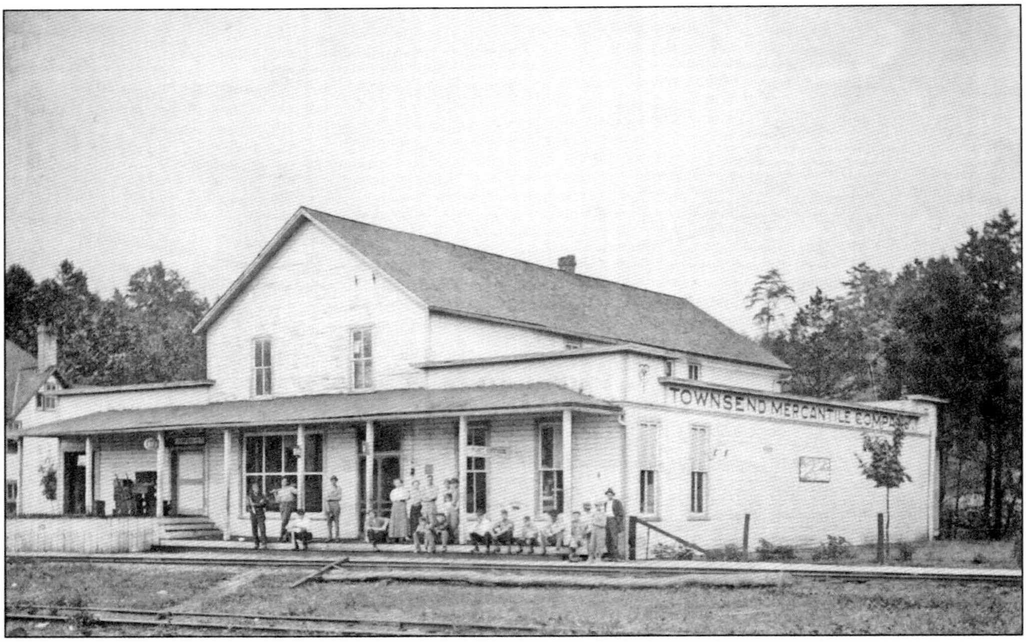

The Little River Lumber Company store was a large building located near the railroad tracks. The office was located on one side of the store, with Townsend Hotel on the other side. The company store was often a gathering place for local residents. Dry goods such as bolts of cloth, pins and needles, thread ribbon, buttons, undergarments, overalls, hats, and shoes were sold in the store. Often, essential items such as rifles, pistols, ammunition, lanterns, ropes, crockery, pots, and pans were also sold. (Above, courtesy of Mary Thompson Headrick; below, courtesy of McNiell Family Collection.)

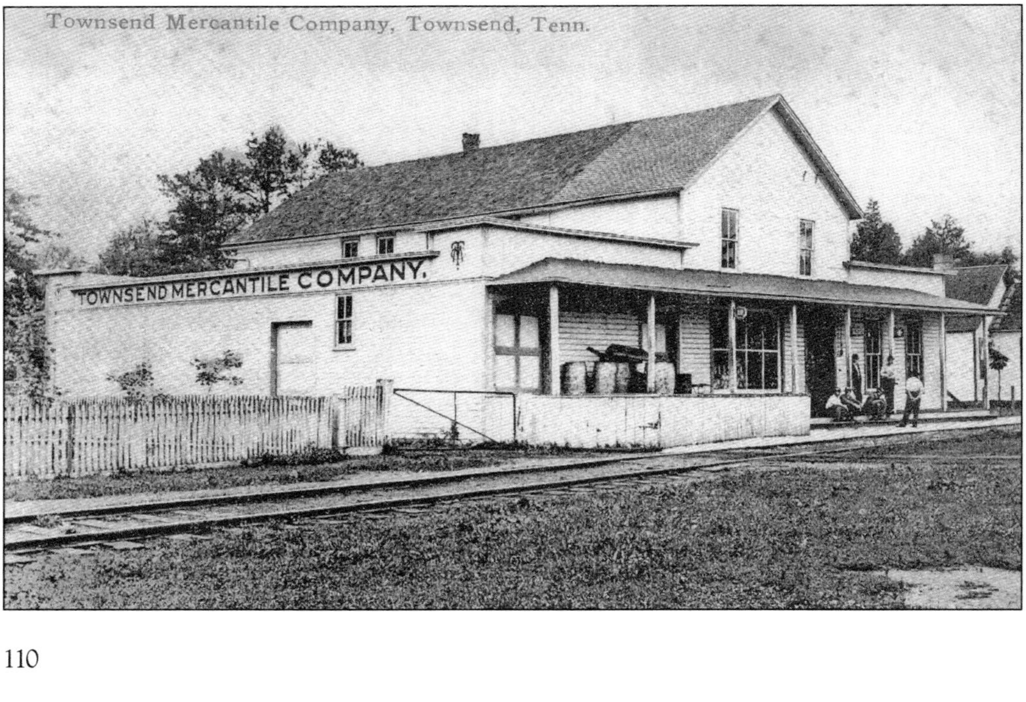

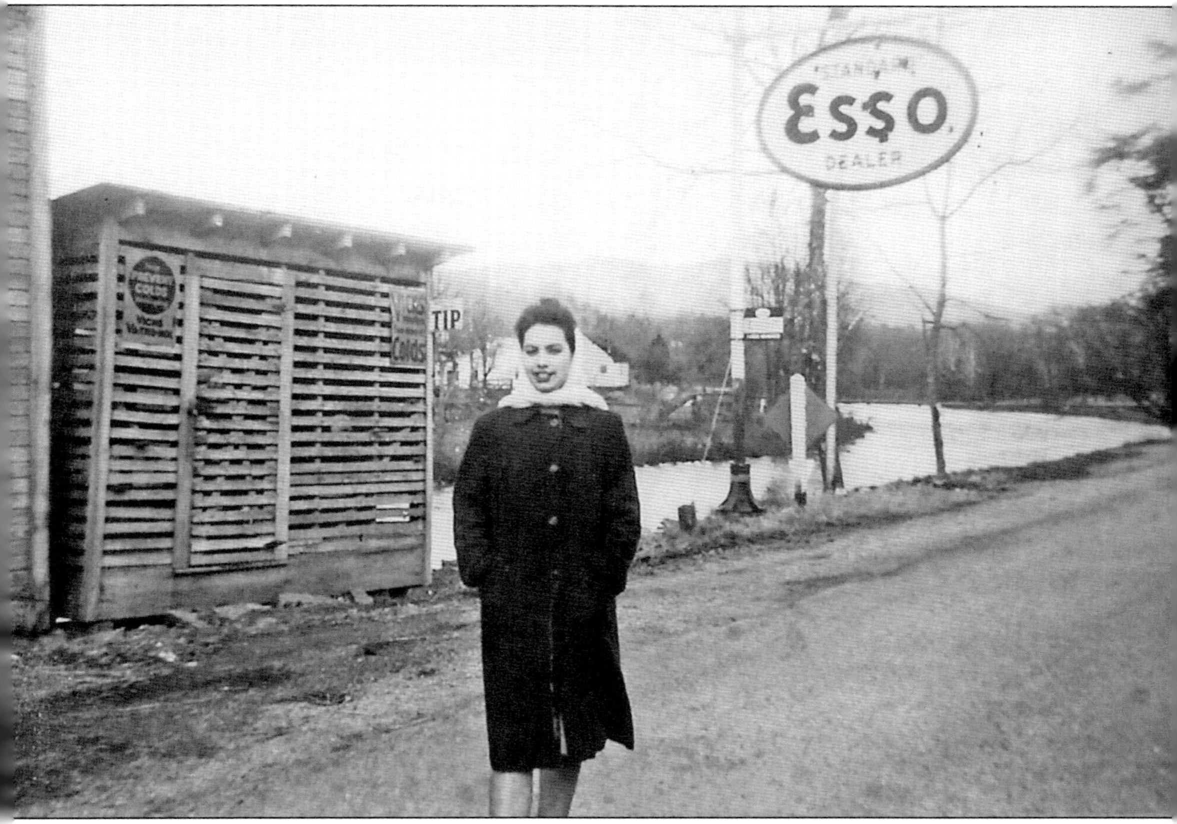

Barbara McNiell Handley stands in front of the grated building used to keep chickens confined when they were brought to the store to be sold or bartered. The store also had apothecary sections with patent medicines, remedies, soaps, castor oils, Epsom salts, iodine, aspirin, catnip tea, and many other items. Workers for the lumber company and railroad were paid in scrip, a term for any substitute for legal tender, often a form of credit. This form of payment can be used in situations where regular money is unavailable, such as remote lumber companies, railroads, or occupied countries in wartime. (Courtesy of McNiell Family Collection.)

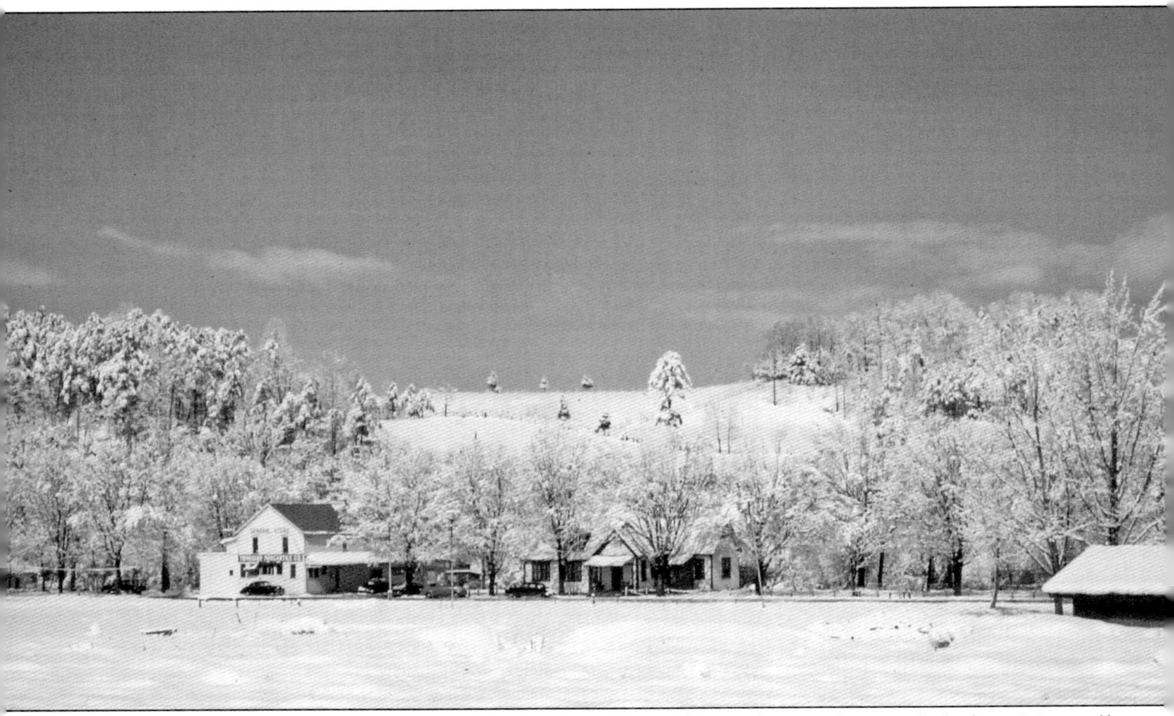

The post office was also located in the company store. The mail arrived twice daily from Maryville and could be picked up at the post office or delivered once a day by Rural Free Delivery. Waiting for the mail in the company store was a great time to congregate and enjoy neighborly fellowship. (Courtesy of the William Derris Slide Collection, MS.2123. University of Tennessee Libraries, Knoxville, Special Collections.)

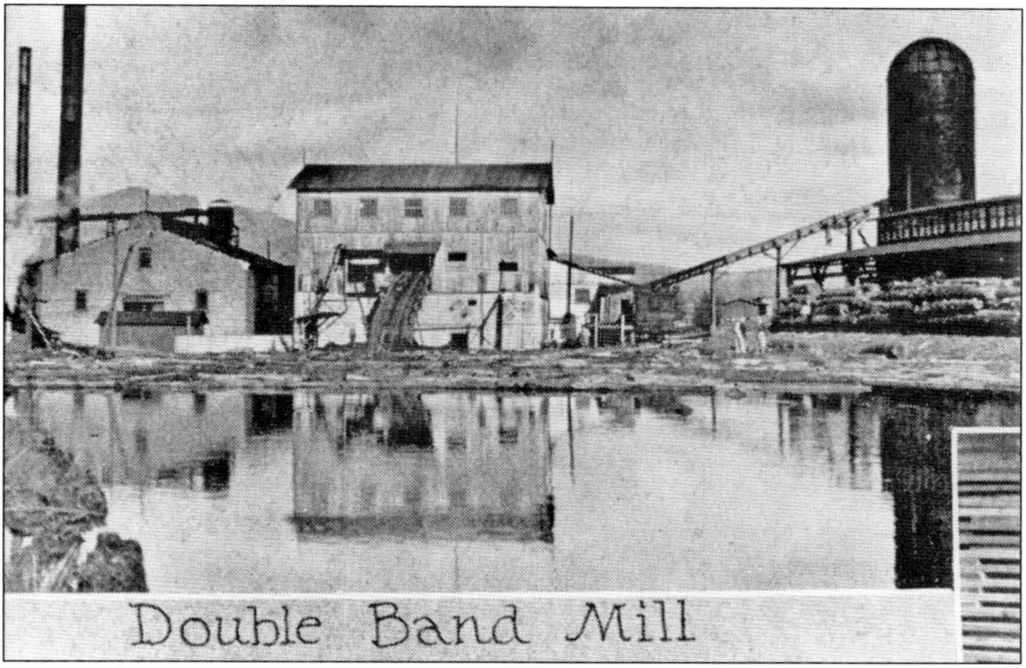

Logs were dumped into a millpond to keep them from drying out before they were milled. Finished lumber was transported on elevated tramways to the drying stacks. The main line of the mill ran east to west of the river, with the mill lying to the south. A sawmill building, steam power plants, and a planing mill were used to produce finished lumber. Blades were mounted on wheels with a diameter large enough to not cause metal fatigue due to flexing when the blade repeatedly changed from a circular to a straight profile. (Courtesy of Stephen Webber, John W. Oliver collection.)

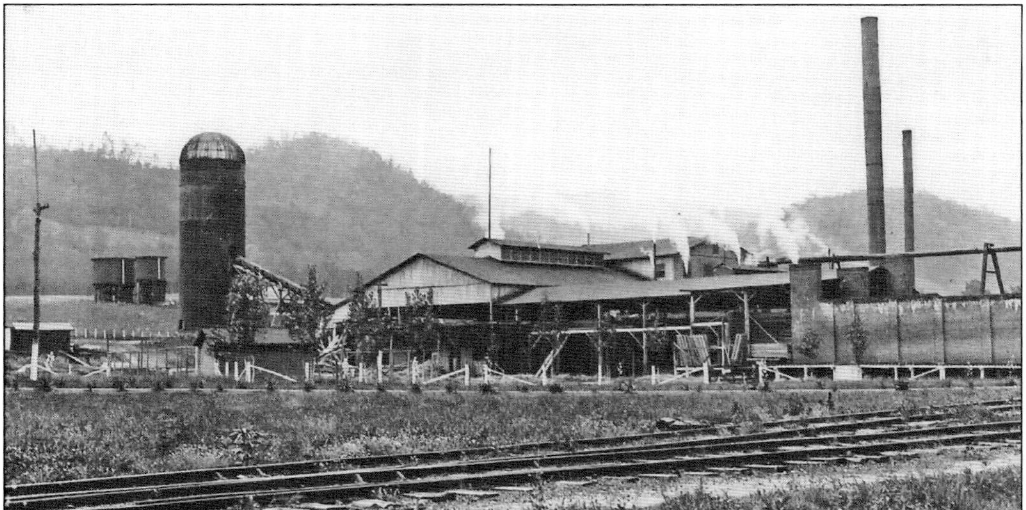

Cured lumber was loaded into boxcars and shipped eight miles to Walland by the Knoxville & Augusta Railway. Once the boxcars were loaded with lumber, they were sealed with a metal strip. Townsend's Little River Lumber Company began to prosper. (Courtesy of McNiell Family Collection.)

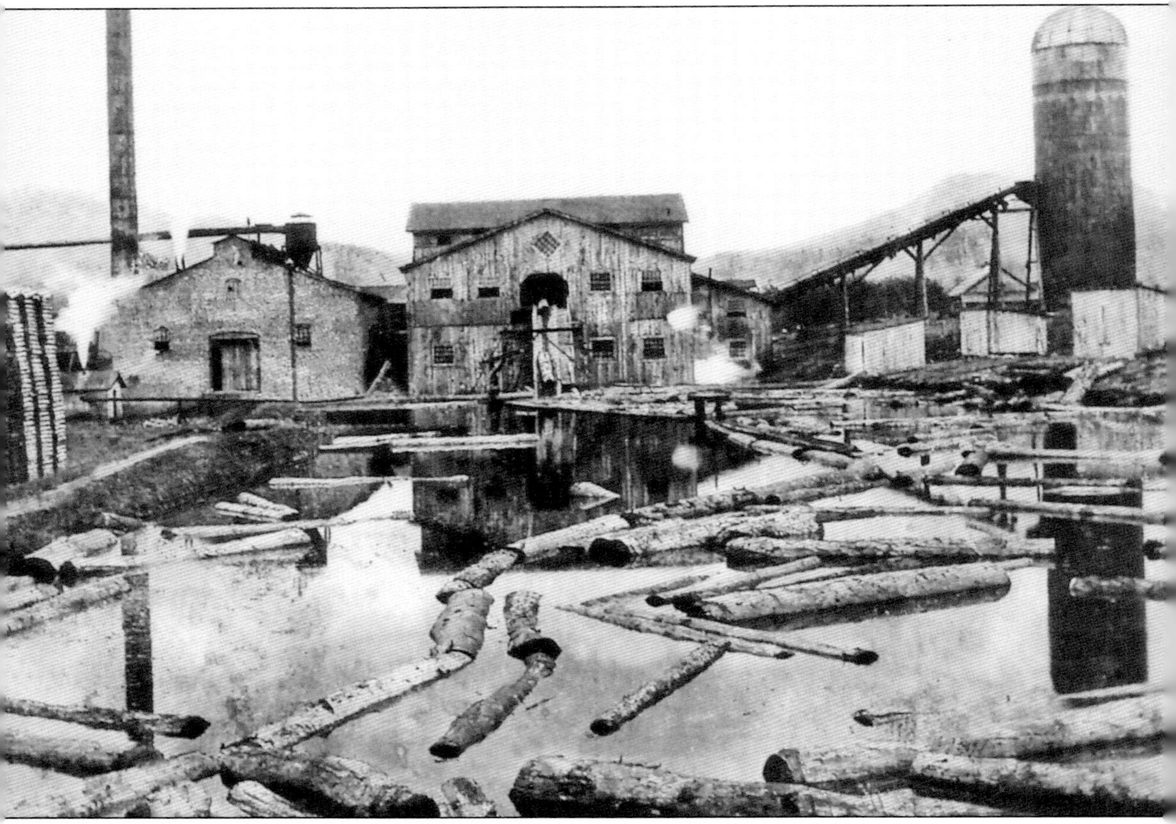

Stacks of freshly cut lumber filled the drying yard next to the pond. Other structures included a burner, planing mill, band sawmill, ice plant, shops, power plant, and water-storage tank. (Courtesy of Missy Tipton Green Photograph Collection.)

LITTLE RIVER RAILROAD.

B. TOWNSEND, Prest. & Gen. Mgr., Townsend, Tenn.
H. McCORMICK, Vice-Prest., Williamsport, Pa.
W. WRIGLEY, Sec'y and Treas., Clearfield, Pa.

R. A. HUFFSTETLER, Gen. Fht. & Pas. Agt Townsend, Tenn
J. P. MURPHY, Supt. & Pur. Agt., "
G. B. TOWNSEND, Chf. Engr., "
D. H. TIPTON, Auditor, "
ENOS COULTER, Car Acct., "

	10	6	4-14	2	Ms.	*June* 23, 1917.	1	9	3	7	11
..	A M	A M	P M	A M	....	(*K. & A. R.R.*)	A M	A M	P M	P M	P M
1	§8 00	§7 30	*3 00	†7 30	....	lve. Knoxville arr.	10 30	10 30	6 00	6 00	8 3
35	§9 05	§9 00	*4 35	†9 00	0	.... Walland [1] ...	8 50	8 50	4 25	4 25	7 2
40	– –	9 05	4 40	9 05	1.7	..... Walkers.....	8 45	8 45	4 20	4 20	– –
48	9 15	9 13	4 48	9 13	5.0	.... Sunshine	8 37	8 37	4 12	4 12	7 1
51	– –	9 16	4 51	9 16	5.2	.... Riverside	8 34	8 34	4 09	4 09	– –
57	– –	9 22	s –	– –	7.0	.. Tuckaleechee ..	s –	– –	– –	4 03	– –
00	9 30	9 25	5 05	9 30	8.0	.... Townsend....	¶8 25	8 25	4 00	§4 00	7 0
M	9 37	A M	5 15	9 37	11.0	...... Forks......	A M	8 13	3 38	P M	6 5
..	10 05	.....	5 45	10 05	19.0	.. Line Springs ..	.....	7 50	3 10	.....	6 2
..	10 25	.....	6 07	10 25	25.4	.Wonderland Park.	.....	7 30	2 50	.....	6 0
..	10 27	.....	6 12	10 27	26.0	.... Elkmont....	.....	7 27	2 47	.....	6 0
..	10 30	.....	6 15	10 30	26.2	.Appalachian Club.	.....	a7 25	†2 45	.....	§6 0
..	A M	.....	P M	A M	....	ARRIVE] [LEAVE	.....	A M	P M	.....	P M

* Daily; † daily, except Sunday; § Sunday only; ¶ daily, except Mon
y; a Monday only; s stops Sunday only.
Connection.—[1] With Knoxville & Augusta Ry. *Central time.*

In 1902, construction of the railroad began in Walland, where it interchanged with the Knoxville & Augusta, predecessor of the Knoxville & Charleston Railway. The eight miles from Walland to Townsend opened for operation on January 1, 1903; the three miles between Townsend and the Forks of Little River were completed in March 1903. (Courtesy of Missy Tipton Green Photograph Collection.)

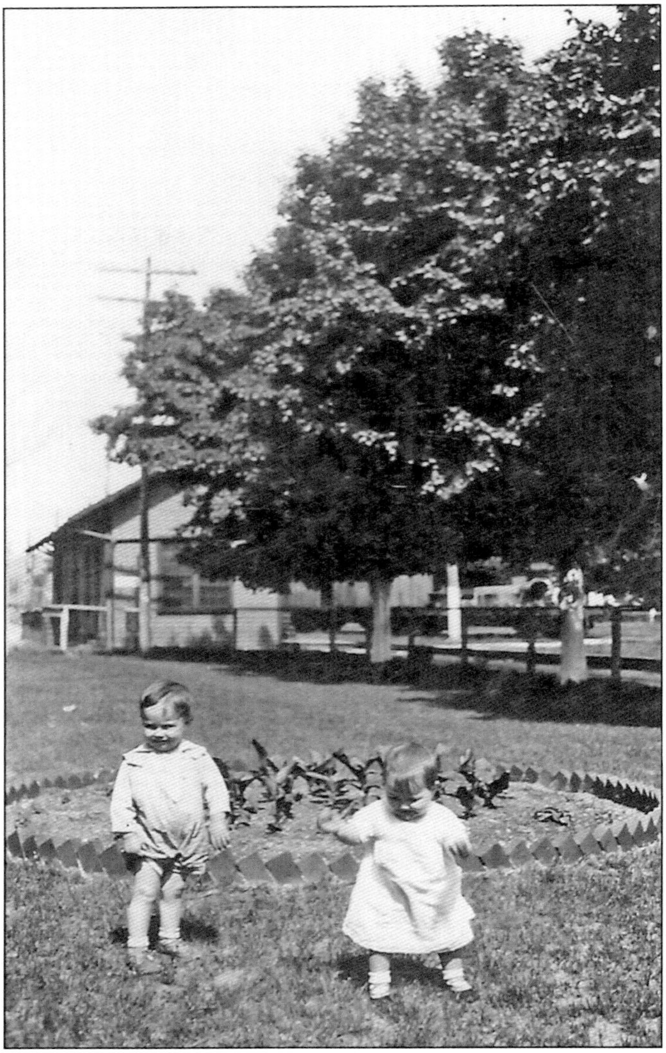

Stuart Jr. (left) and Bobbie McNiell, children of Stuart Sr. and Stella Myers McNiell, are pictured at the Little River Lumber office in 1928. Stuart McNiell Sr. was a bookkeeper for the Little River Lumber Company. The office building had a white board exterior with dark green trim, and a porch leading from the back had dark green benches on each side. The office had a door leading from the railroad side and one from the side by the river. The switchboard, which was in a small room on the left when entering from the railroad side, was used to connect to the outside world along with phones in key personnel homes and some company buildings. Also inside the office building, as one ventured into the hall, was a walk-in safe. On the right side of the hall were two offices, which were for W.B. Townsend and his secretary. Later, the offices were for D.H. Tipton and his secretary. On the left side of the hall was a large office, which housed the bookkeeping department. Exiting from this room led to two offices for other leaders of the company. In the aforementioned large room, Stuart McNiell Sr. kept the books for Little River Railroad & Lumber Company and Townsend Light Company. On payday, the men lined up outside his window to the hall to receive their pay. Some men only received $1 a day for their work. As time passed, salaries increased, but they could never keep pace with wages offered by the aluminum company. (Courtesy of McNiell Family Collection.)

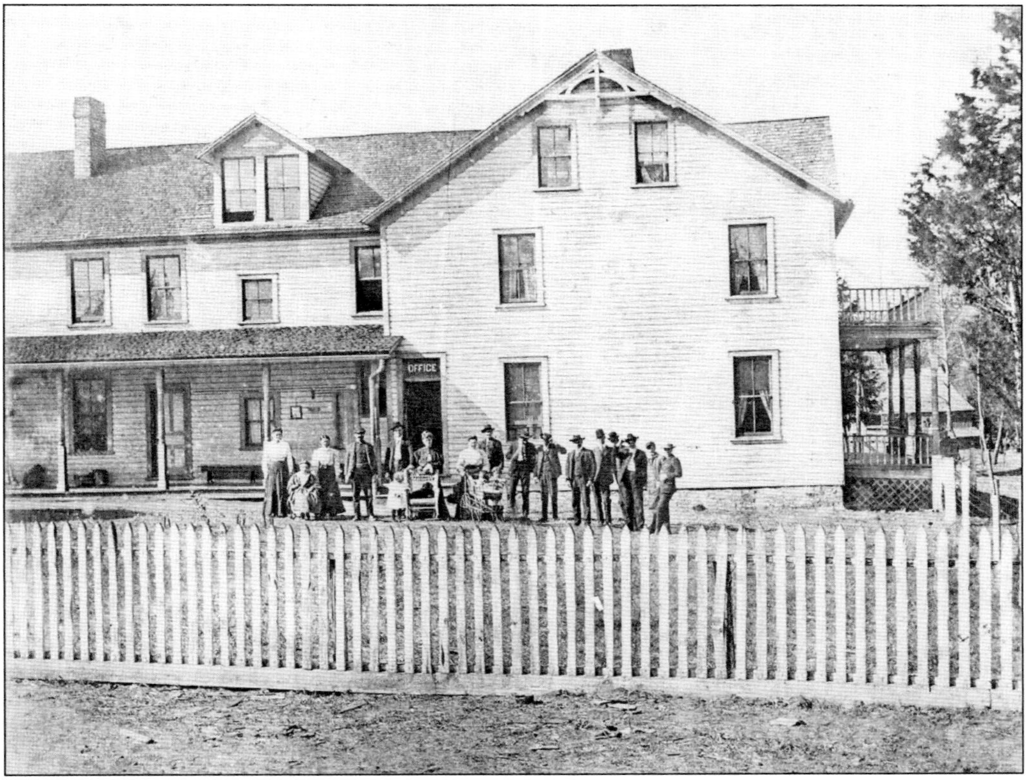

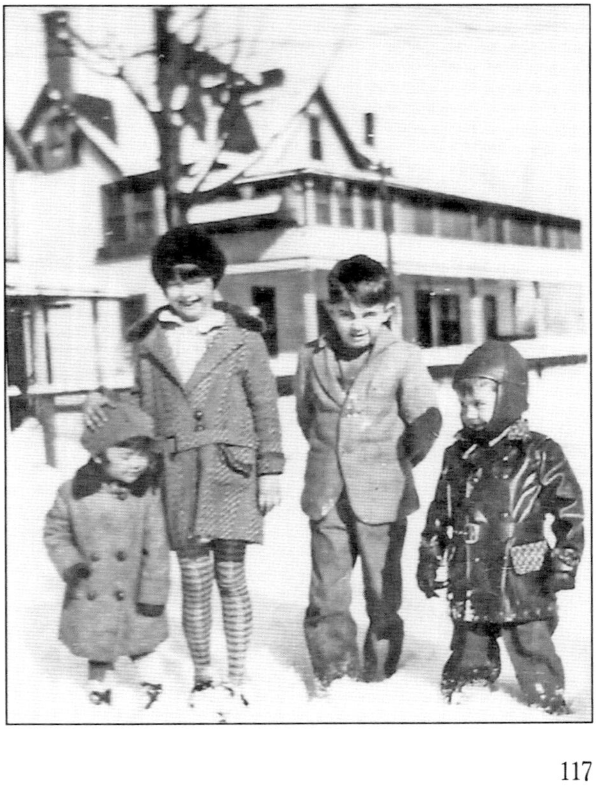

Pictured above is the hotel that was named after W.B. Townsend. It was three-stories tall and offered rooms for short-term or long-term rental. The back of the hotel, which faced the road, had a small porch that offered guests a good view of the river. Pictured at right are, from left to right, Barbara McNiell, Micky Townsend, Frank Bud Townsend, and Stuart McNiell at the Townsend Hotel in 1931. (Courtesy of McNiell Family Collection.)

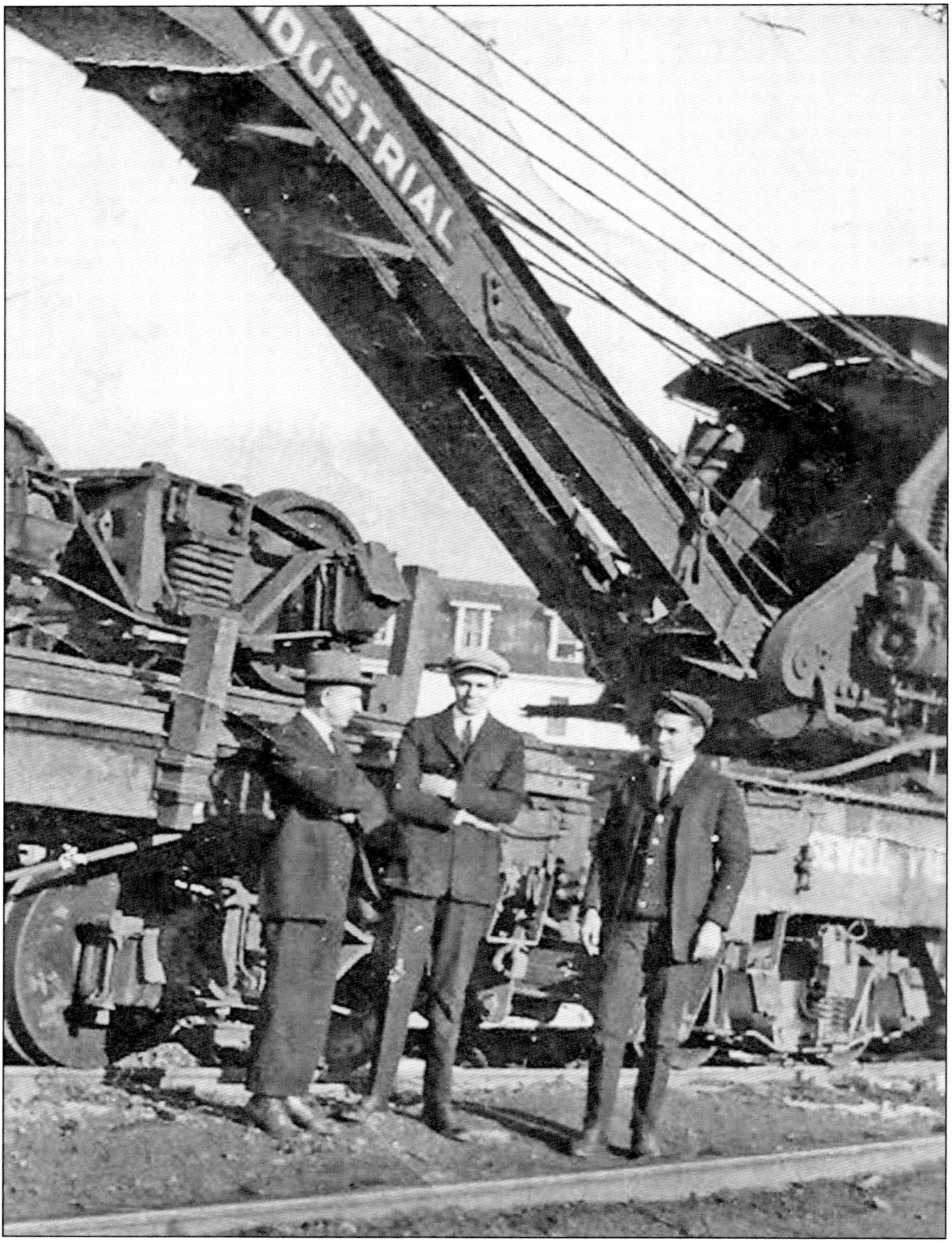

From left to right, Leonard Myers, Clyde Headrick and Lee Myers stand in front of the Little River Lumber Company crane. The crane picked up logs after they were brought off the mountain, and it was used to unload the logs at the mill pond at Townsend. Little River Lumber Company also used a crane or log loader to move its portable houses to a new logging camp. The houses were set on flat cars by the crane or loader and pulled by a Shay engine to the new camp. (Courtesy of Marilyn Myers Byrd Photograph Collection).

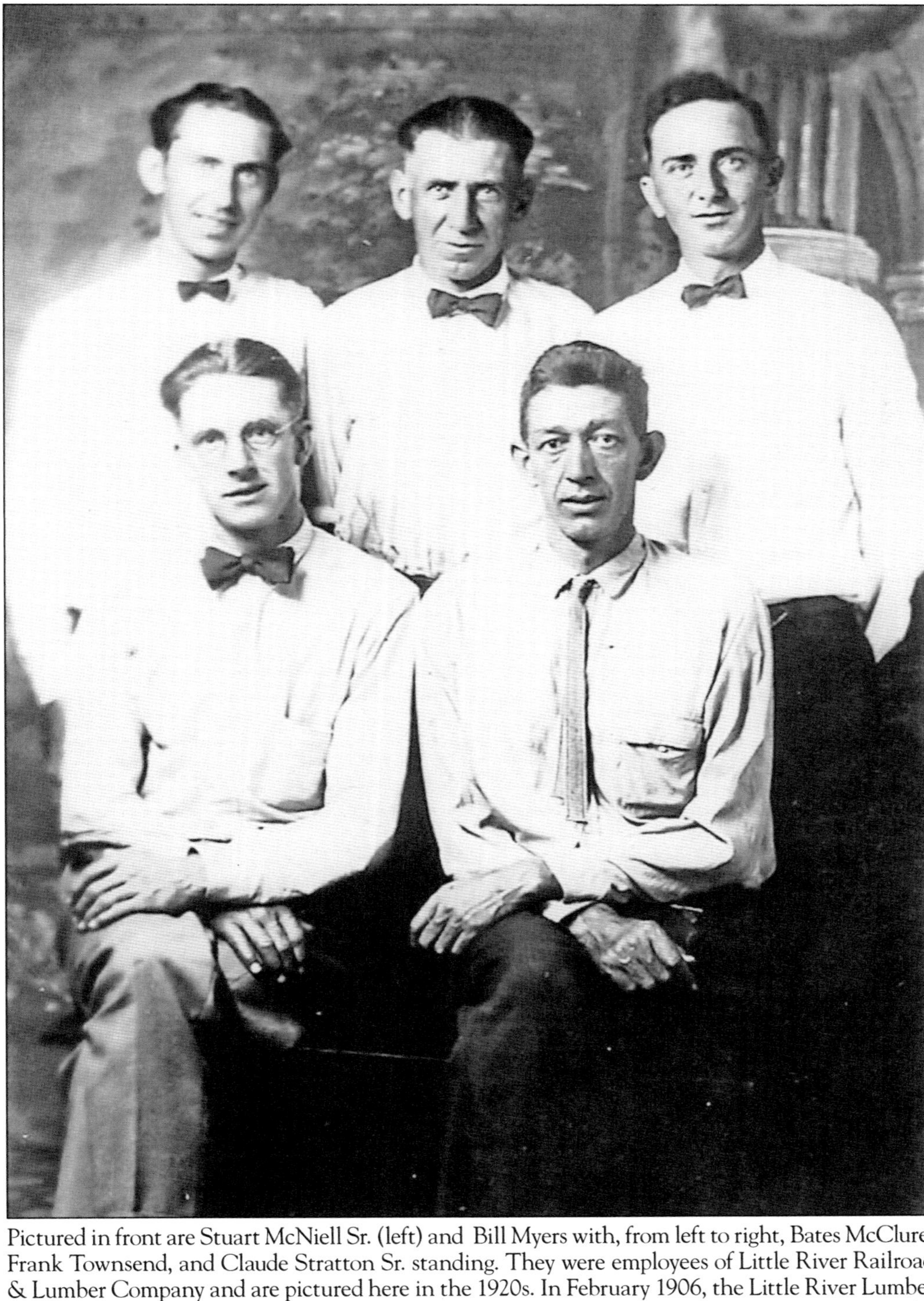

Pictured in front are Stuart McNiell Sr. (left) and Bill Myers with, from left to right, Bates McClure, Frank Townsend, and Claude Stratton Sr. standing. They were employees of Little River Railroad & Lumber Company and are pictured here in the 1920s. In February 1906, the Little River Lumber Company mill burned to the ground, but it was immediately rebuilt to be bigger and better than the original. In June 1916, this new mill burned, and again the Little River Lumber Company rebuilt. This mill lasted until the end of operations, around 1941. (Courtesy of McNiell Family Collection.)

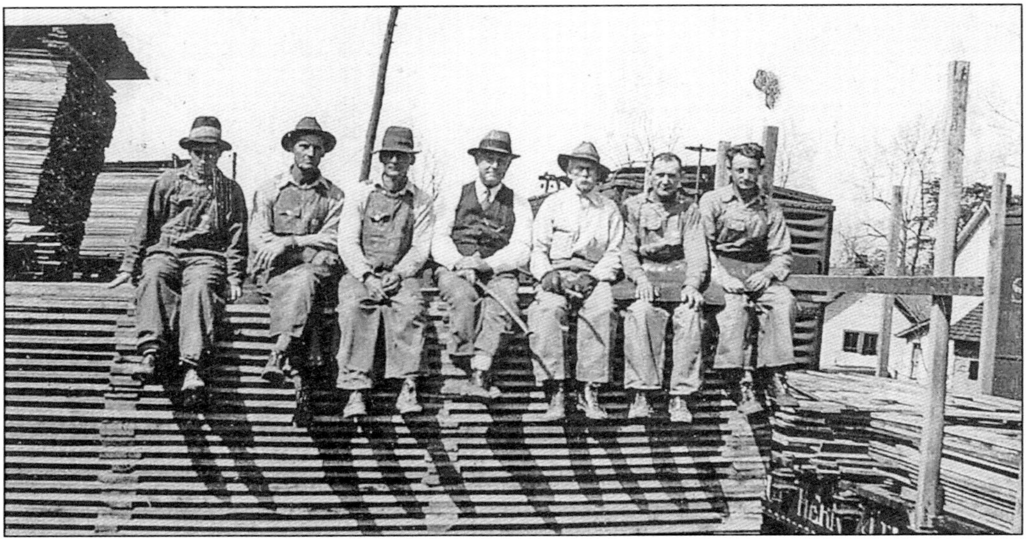

Residents of Townsend used their talents and skills to provide goods and services for their community and to meet the demands of the expanding building industry. Townsend had to transition from a large farming community into an industrial community, which required strong work ethic, skills, and community support. (Courtesy of Missy Tipton Green Photograph Collection.)

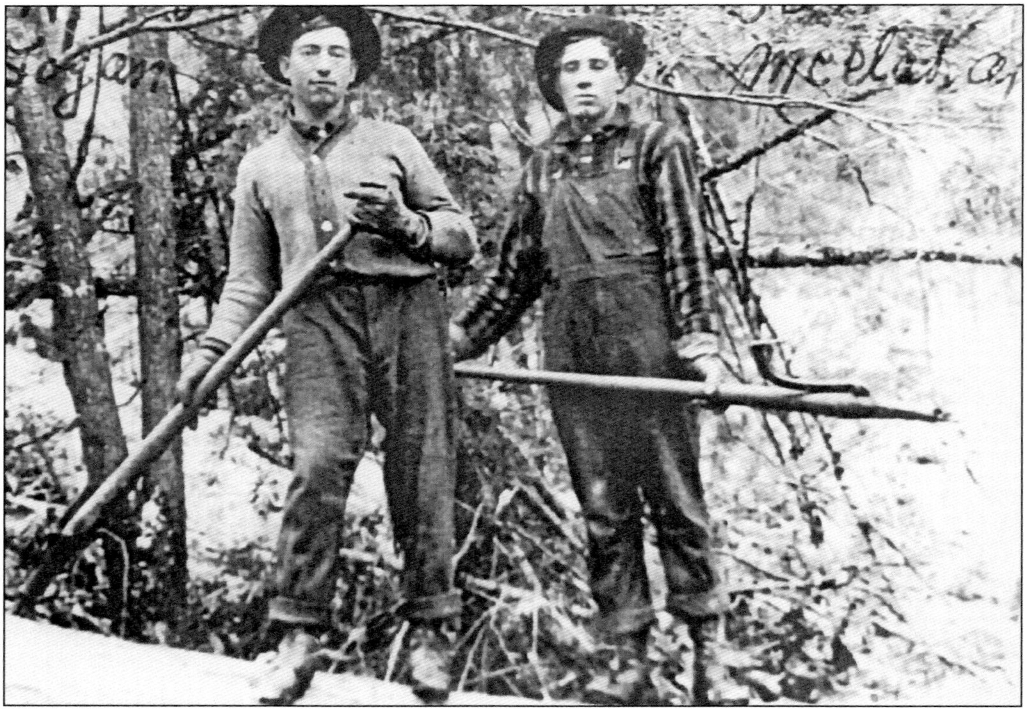

Sam McClanahan (right) and Henry Logan are pictured here working for Little River Lumber Company. Sam was just a boy when the railroad and lumber business started in Townsend. It was common for people who lived close to the lumber company and mill to begin working there at a young age. Sam worked his way up to a fireman on the locomotives; he was responsible for tending to the fire used to run the steam engine. (Courtesy of Eddie McClanahan.)

Pictured is a page of the Little River Railroad ledger book. The logging operations of Little River Lumber Company consisted of different phases. After the tree was found, it was cut with a two-man crosscut saw. The moving of the fallen timber came next, and logs were taken to the railhead. The last phase was loading the logs onto the empty boxcars for transportation back to Townsend. (Courtesy of McNiell Family Collection.)

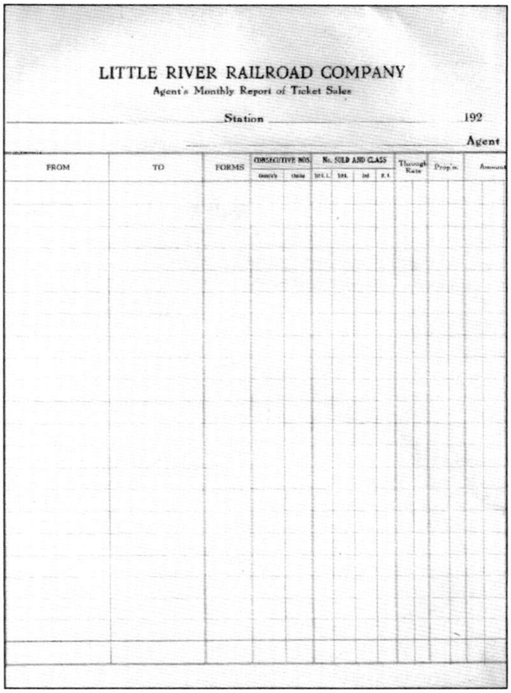

LITTLE RIVER RAILROAD COMPANY

Agent's Monthly Report of Ticket Sales

Station 192

Agent

FROM	TO	FORMS	CONSECUTIVE NOS.		No. SOLD AND CLASS				Through Rate	Prop'n.	Amount

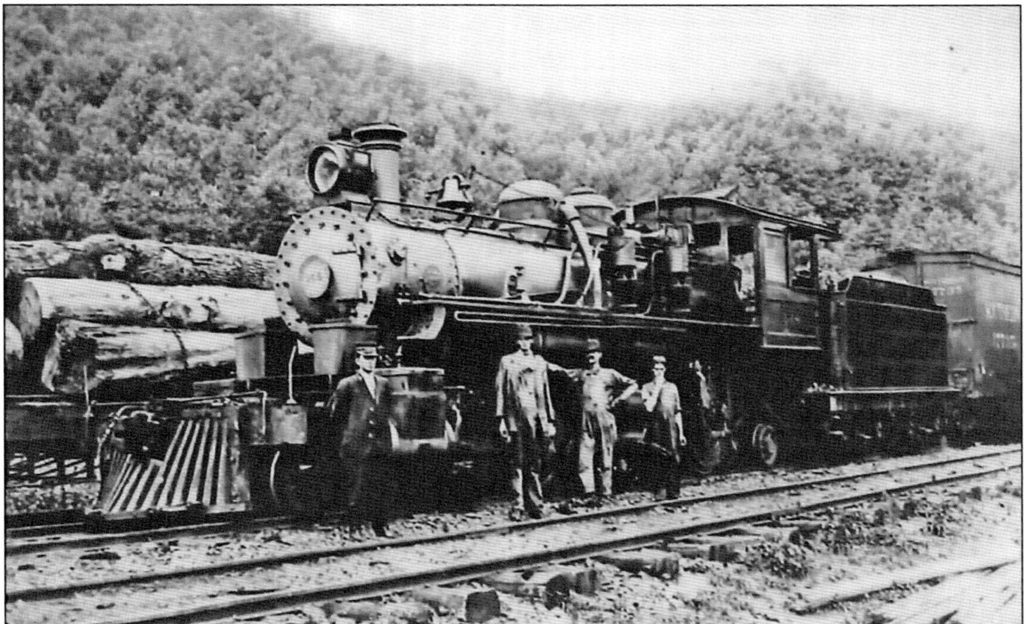

During the Little River Lumber Company and Railroad era, the railroads could either be a standard or a narrow-gauge rail. Standard could have a distance of 4 feet and 8.5 inches between the rails. The narrow gauge was only 3 feet between the rails but was cheaper to build and easier to climb the mountain. W.B. Townsend built most of his as a standard gauge so he would not have to transfer his goods from one car to another for shipment. (Courtesy of Great Smoky Mountains National Park.)

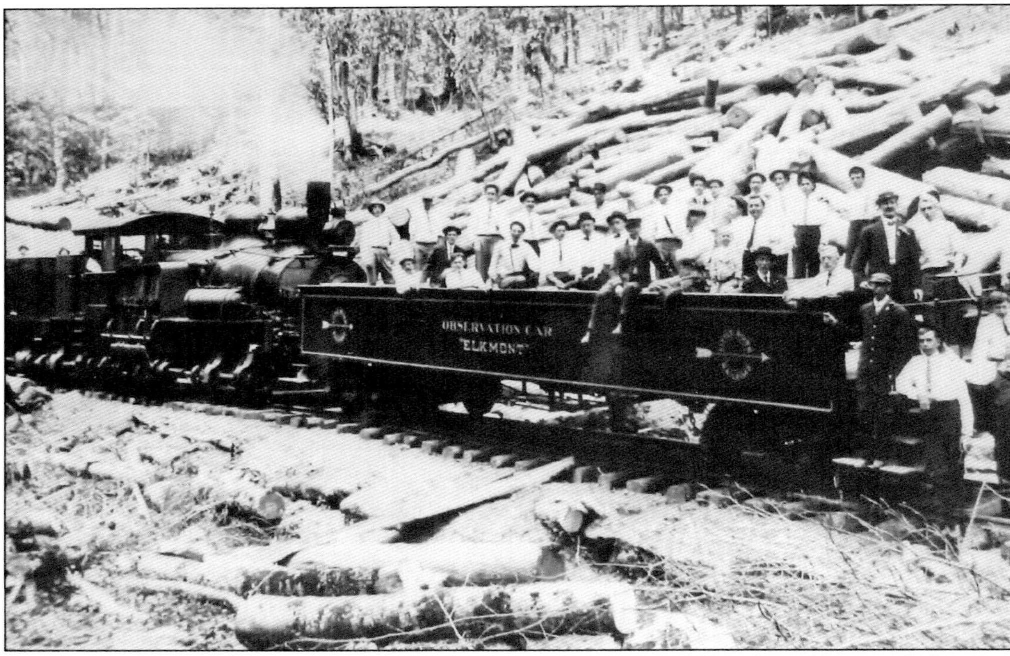

On July 2, 1909, it was announced in a newspaper that Little River Railroad would start a daily service from Knoxville's Southern station to Elkmont. A Sunday-only excursion had already been started, but the number of visitors going to Elkmont had increased dramatically. The Sunday-only "Elkmont Special" was a joint effort with Little River Railroad and Southern Railroad. It was a through train running from Knoxville with no change of cars, but the engine changed over at Walland before continuing on to Elkmont. (Courtesy of Great Smoky Mountains National Park.)

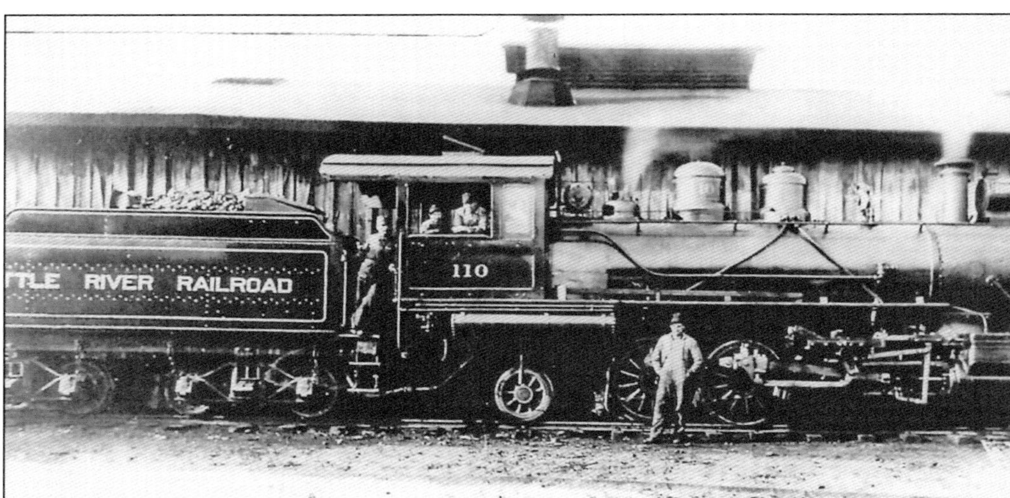

In this image from the 1920s, Jack Foster stands in the doorway of Little River Engine No. 110, the brakeman is Luther Swann, and J.K. Foster stands at right. This engine was known as "Mrs. Townsend's Engine"; it carried a portrait of Margaret Townsend, W.B.'s first wife, in the cab window. This engine pulled Margaret Townsend's funeral train. (Courtesy of Missy Tipton Green Photograph Collection.)

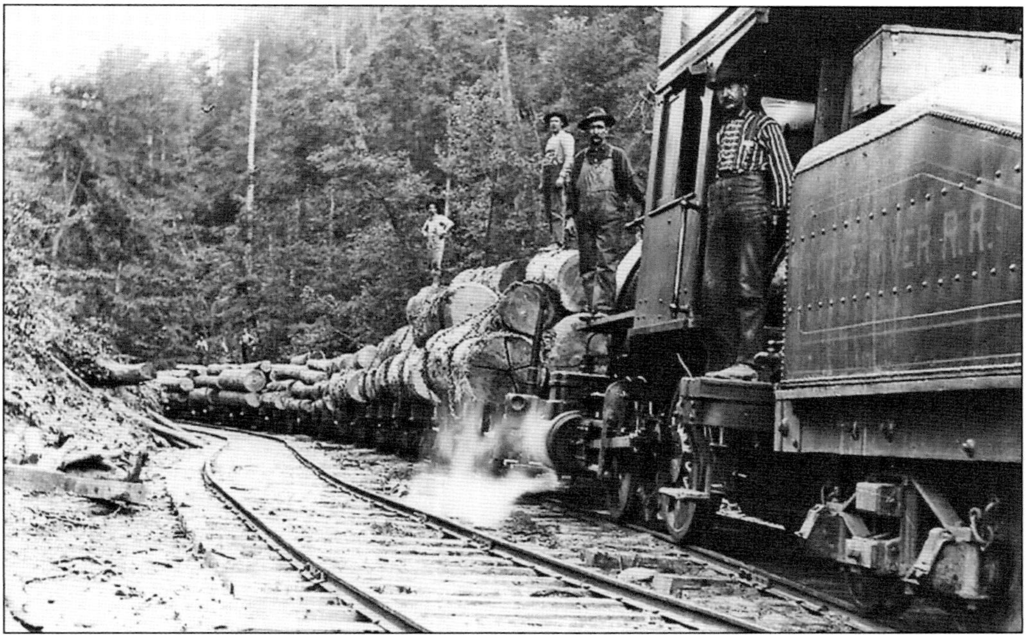

This is Little River Railroads first rod Engine No. 1. These logs had been brought out of West Prong and Laurel Branch by engines that could go around sharp curves and steep grades. They were then loaded into Engine No. 1 and transported back to Townsend. (Courtesy of McNiell Family Collection.)

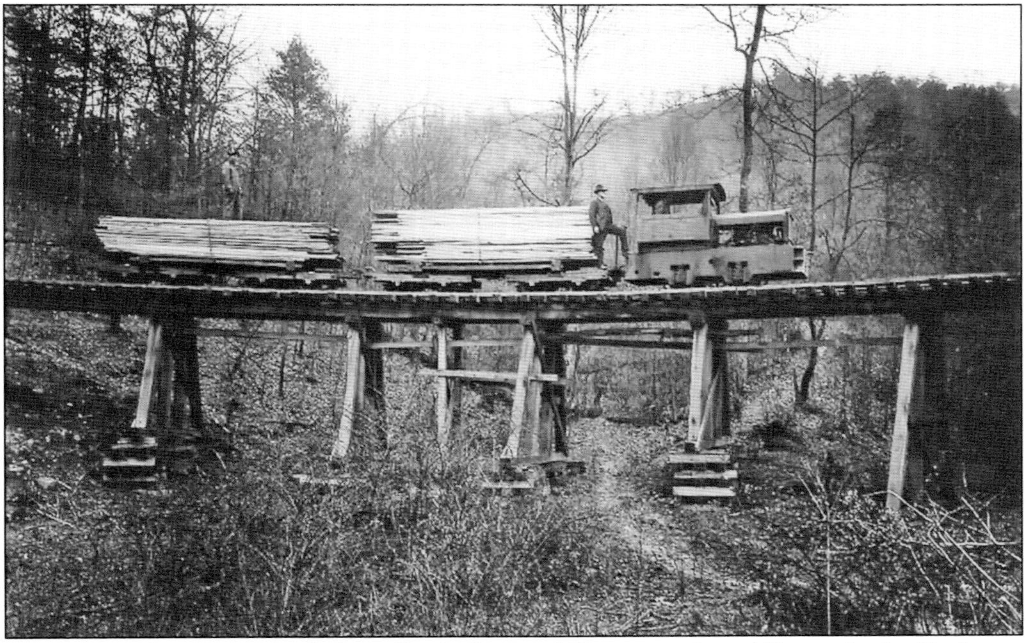

After the lumber went through the mill, it had many different uses. It could be used to build a home, store, or other structures. If the distance the lumber was being transported was not very far, it could be hauled by a much smaller engine. (Courtesy of Great Smoky Mountains National Park.)

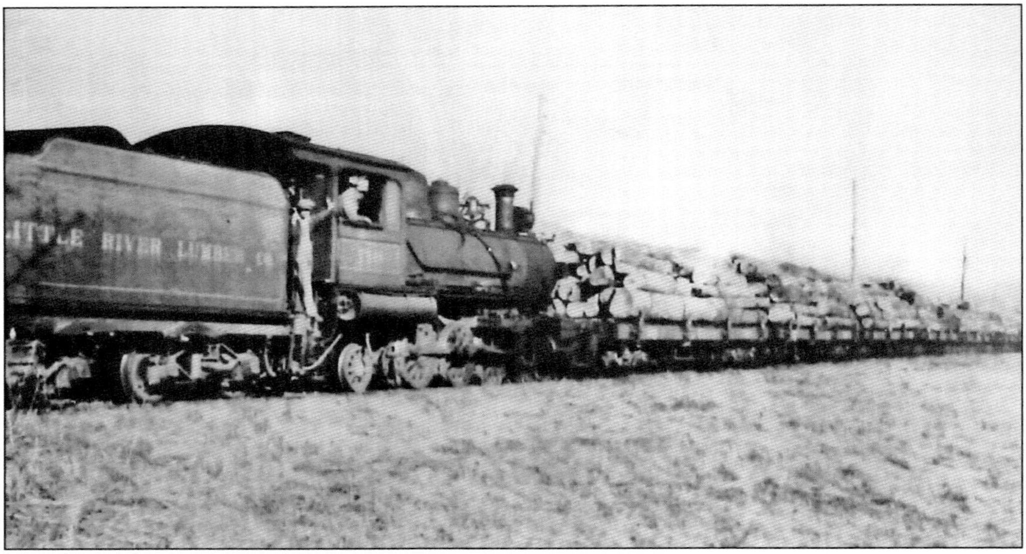

It was often said that Engine No. 110 could have a large load of logs and still move 35 or 40 miles per hours. The railroad workers used to say the engine "was talking to them" as it chugged up the mountain. This engine was hard working and many engineers enjoyed riding it. Woody Dew was one of the engineers on Engine No. 110 for many years. (Courtesy of Great Smoky Mountains National Park.)

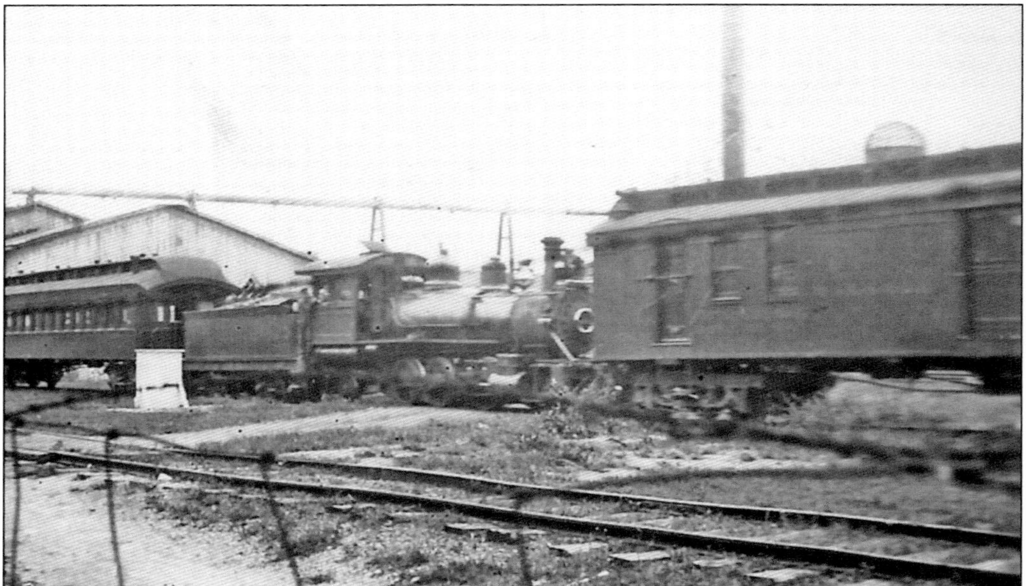

Accidents happened often at Little River Railroad & Lumber Company; the workers' jobs were very dangerous. If there were a fatality among the workers or if one of the workers passed on, the funeral train was pulled by Engine No. 110. The train took the mourners to the funeral and, after the service, carried them back up the river to their shanty houses or jobs. The officials of the Little River Lumber Mill thought the train ride was statelier than a horse and wagon. (Courtesy of Great Smoky Mountains National Park.)

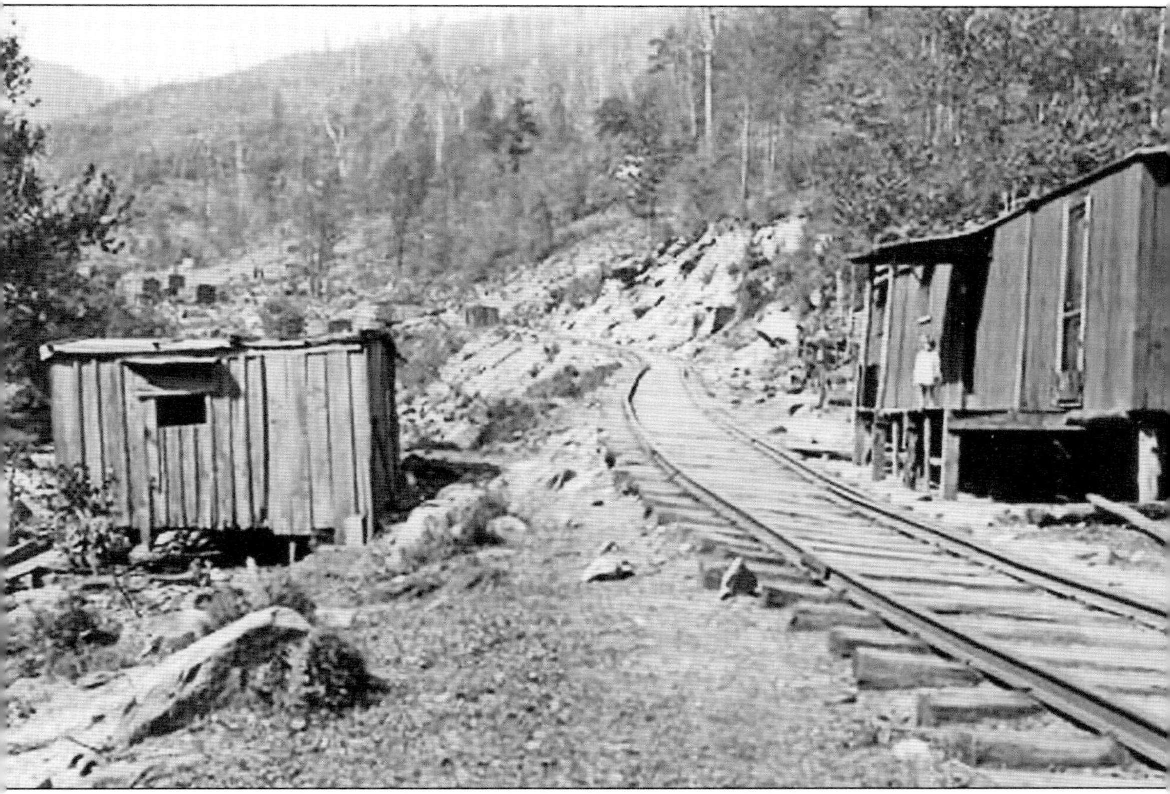

Pictured are setoff houses above the Tremont area. Setoff homes were exactly that—homes set off from the train tracks. This allowed for easy movement when logging was finished in an area and camp needed to be moved elsewhere. Large families would increase the size of their homes by putting two of three of these buildings together. (Courtesy of Little River Railroad Museum Photograph Collection.)

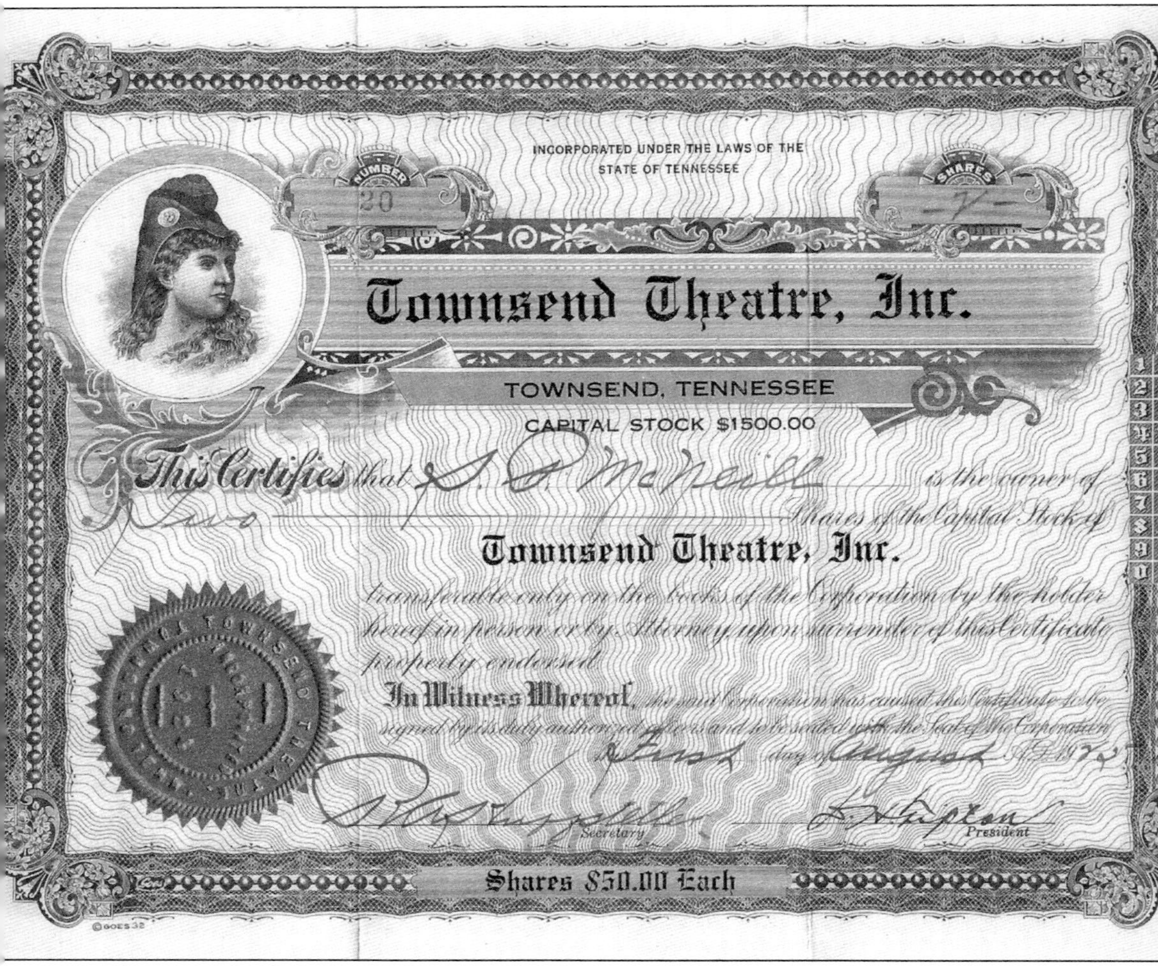

INCORPORATED UNDER THE LAWS OF THE STATE OF TENNESSEE

NUMBER 20 SHARES

Townsend Theatre, Inc.

TOWNSEND, TENNESSEE

CAPITAL STOCK $1500.00

This Certifies that S. P. McNeill is the owner of Two Shares of the Capital Stock of

Townsend Theatre, Inc.

transferable only on the books of the Corporation by the holder hereof in person or by Attorney upon surrender of this Certificate properly endorsed.

In Witness Whereof, the said Corporation has caused this Certificate to be signed by its duly authorized officers and to be sealed with the Seal of the Corporation this First day of August A.D.

Secretary President

Shares $50.00 Each

This is a capital stock certificate for the Townsend Theatre, Inc., whose owner was Stuart P. McNiell. In 1922, Stuart P. McNiell Sr. went to work as bookkeeper for Little River Lumber Company and Little River Railroad. A theater was built in the area of the lumber company for the employees and their families to enjoy. In 1936, Stuart was elected superintendent when D.H. Tipton was elected president after the death of W.B. Townsend. He acted in this role until the closure of Little River Lumber Company in 1939 and Little River Railroad in 1940 when it was closed out. As superintendent, McNiell oversaw operations of the bookkeeping department, lumber yard, mill, woods department, and shop. Little River Light and Power Company had a powerplant at Townsend by 1904, and McNiell served as bookkeeper for this company as well. (Courtesy McNiell Family Photograph Collection.)

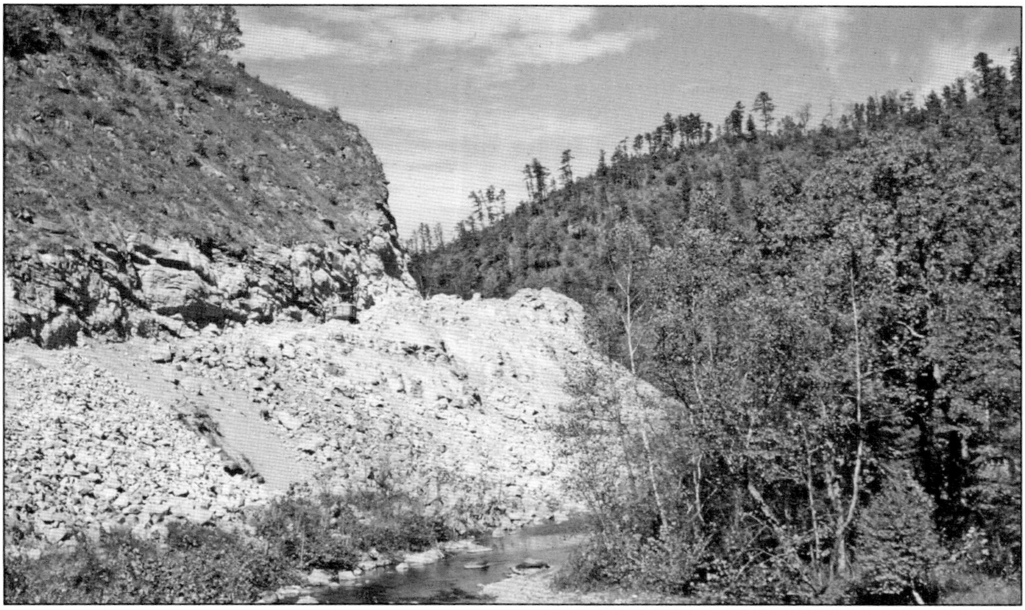

This is the road to Townsend in the late 1950s or early 1960s. It took a lot of digging and blasting to get this road, now Highway 321, cut out of the mountain. The early road into Tuckaleechee Cove was located on the other side of the river. (Courtesy of the William Derris Slide Collection, MS.2123. University of Tennessee Libraries, Knoxville, Special Collections.)

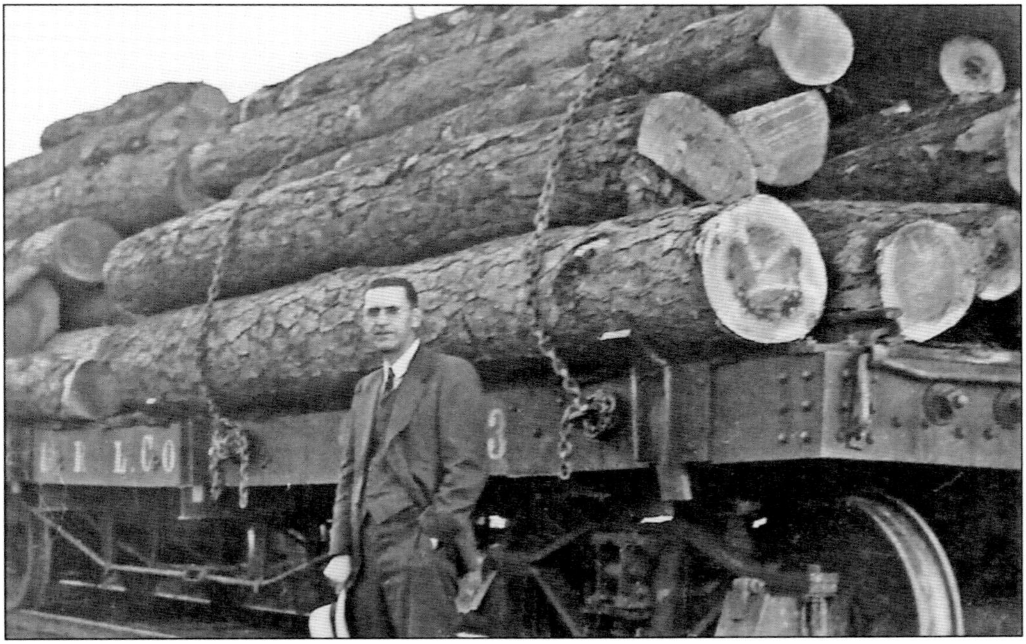

Stuart McNiell Sr. stands beside a load of the last logs cut by Little River Lumber Company in 1938. Stuart came to work for Little River Lumber Company in the 1920s and remained until its closure. He met his wife, Stella Myers, when he came to Townsend. He worked many years as the bookkeeper for Little River Railroad & Lumber Company, and after D.H. Tipton retired, he became the superintendent. (Courtesy of McNiell Family Collection.)